ADVANCES IN
ORGAN BIOLOGY

Volume 1 • 1996

PREGNANCY AND PARTURITION

ADVANCES IN ORGAN BIOLOGY

PREGNANCY AND PARTURITION

Series Editor: E. EDWARD BITTAR
Department of Physiology
University of Wisconsin

Guest Editor: TAMAS ZAKAR
Perinatal Research Centre
University of Alberta

VOLUME 1 • 1996

JAI PRESS INC.

Greenwich, Connecticut *London, England*

55 Old Post Road No. 2
Greenwich, Connecticut 06836

JAI PRESS LTD.
38 Tavistock Street
Covent Garden
London, England WC2E 7PB

ISBN: 1-55938-639-8

Transferred to digital printing in 2009.

CONTENTS

LIST OF CONTRIBUTORS

James G. Allardice — Department of Obstetrics, Gynecology, and Reproductive Sciences
University of Manitoba
Manitoba, Canada

Michael L. Boroditsky — Department of Obstetrics, Gynecology, and Reproductive Sciences
University of Manitoba
Manitoba, Canada

Jamal Dakour — Signal Transduction Laboratories
Perinatal Research Centre
Department of Medicine
University of Alberta

Charles A. Ducsay — Center for Perinatal Biology
Departments of Physiology and Pediatrics
School of Medicine
Loma Linda University
Loma Linda, California

P. Audrey Fernandes — Department of Obstetrics, Gynecology, and Reproductive Sciences
University of Manitoba
Manitoba, Canada

Frank Hertelendy — Department of Obstetrics and Gynaecology
St. Louis University Health Sciences Center

Jonathan J. Hirst — Department of Physiology
Monash University
Clayton, Victoria, Australia

Hongshi Li	Signal Transduction Laboratories Perinatal Research Centre Departmént of Medicine University of Alberta
John A. McCoshen	Department of Obstetrics, Gynecology, and Reproductive Sciences University of Manitoba Manitoba, Canada
Wendy J. McLaren	Department of Physiology Monash University Clayton, Victoria, Australia
Jane E. Mijovic	Perinatal Research Centre Departments of Physiology, Obstetrics and Gynaecology, and Pediatrics University of Alberta
Bryan F. Mitchell	Perinatal Research Centre Department of Obstetrics and Gynaecology University of Alberta
Miklós Molnár	Institute of Pathophysiology Semmelweis University Medical School Budapest, Hungary
Donald W. Morrish	Signal Transduction Laboratories Perinatal Research Centre Department of Medicine University of Alberta
David M. Olson	Perinatal Research Centre Departments of Physiology, Obstetrics and Gynaecology, and Pediatrics University of Alberta
Joon W. Rhee	Center for Perinatal Biology Departments of Physiology and Pediatrics School of Medicine Loma Linda University Loma Linda, California

Gregory E. Rice	Department of Physiology Monash University Clayton, Victoria, Australia
Kiyoshi Shimada	Department of Animal Physiology School of Agricultural Sciences Nagaya University Nagaya, Japan
Kousaku Tanaka	Laboratory of Animal Breeding Faculty of Agriculture Kyushu University Fukuoka, Japan
Geoffrey D. Thorburn	Department of Obstetrics and Gynaecology Monash University Clayton, Victoria, Australia
Miklós Tóth	First Department of Biochemistry Semmelweis University Medical School Budapest, Hungary
I. Ross Young	Department of Physiology Monash University Clayton, Victoria, Australia
Tamas Zakar	Perinatal Research Centre Department of Obstetrics and Gynaecology University of Alberta

PREFACE

One of the mysteries of mammalian reproduction is the physiologic process that determines the length of gestation. The proper timing of birth ensures that the young individual is sufficiently developed to survive and adapt in the extrauterine environment, and that the mother is capable to provide nutrition and protection to the newborn. This volume summarizes new knowledge obtained by many researchers seeking to unravel the complex mechanisms that contribute to the maintenance and termination of pregnancy. The most important common goal of these efforts is to reduce the incidence of preterm birth, which is the leading cause of perinatal morbidity and mortality in numerous countries.

Separate chapters are devoted to the best studied animal models of parturition. In sheep, the fetus is in control of the timing of its own birth, while in avian species, oviposition is evidently determined by the female laying the fertilized egg. In humans and non-human primates, the roles of the fetus and the mother are more balanced, and involve a complicated and poorly understood interplay between the mother, the fetus, and the placenta. Some major aspects of these interactions, such as trophoblast function, myometrial contractility, and the endocrine-paracrine systems, are discussed in further chapters.

The authors of this volume are active investigators, performing cutting edge research in perinatology. Thanks are due to them for finding time to write. Further, it would have been impossible to organize the manuscripts into a book without the gratefully acknowledged secretarial help of Ms. Sheila McManus.

Tamas Zakar
Guest Editor

LATE PREGNANCY AND PARTURITION IN THE SHEEP

Wendy J. McLaren, I. Ross Young, and
Gregory E. Rice

Advances in Organ Biology
Volume 1, pages 1–29.

ISBN: 1-55938-639-8

ABSTRACT

Parturition in the sheep is preceded by activation of the fetal hypothalamic-pituitary-adrenal (HPA) axis. Late in the last third of gestation the hypothalmic peptides corticotrophin releasing hormone (CRH) and arginine vasopressin (AVP) act to stimulate the release of adrenocorticotrophic hormone (ACTH) from the fetal pituitary gland. The pre-parturient ACTH surge increases the secretion rate of cortisol from the fetal adrenal gland. Cortisol provides the trigger for labor onset by the induction of changes in placental steroid output. An increase in P_{450} 17α-hydroxylase expression causes a decrease in maternal progesterone concentrations and a concomitant increase in placental estrogen output. The rise in the estrogen/progesterone ratio enhances the sensitivity of the myometrium to stimulatory agonists such as prostaglandin $F_{2\alpha}$ ($PGF_{2\alpha}$), prostaglandin E_2 (PGE_2), and oxytocin. Furthermore, the change in the steroid milieu increases the secretion rate of $PGF_{2\alpha}$ from uterine tissues, including the maternal placenta, endometrium, and fetal membranes. Birth is a complex sequence of endocrine events culminating in coordinated expulsive mechanisms. Vaginal distension in the second stage of labor causes oxytocin release from the maternal posterior pituitary gland (the "Ferguson" reflex). The exact role of oxytocin in the initiation of labor, however, is controversial. Despite high uterine oxytocin receptor concentrations in late pregnancy, oxytocin concentrations in maternal and fetal plasma are not increased until labor onset. Oxytocin acts to enhance $PGF_{2\alpha}$ release in late pregnancy. $PGF_{2\alpha}$ production by the endometrium and maternal cotyledon stimulates synchronous and forceful contractions of the uterus. PGE_2 production by the fetal trophoblast activates the fetal HPA axis. Furthermore, PGE_2 is a major contributor to cervical effacement and softening. Prematurity and postmaturity represent great hazards to the neonate. If the fetus is to survive outside the uterus without being compromised, fetal organ systems, including the respiratory, thermoregulatory, and gastrointestinal systems, must be adequately developed. It seems appropriate then that maturation of body functions essential for postnatal survival are activated by the same mechanism that triggers labor onset. Fetal glucocorticoids promote development and differentiation of fetal organ systems. Cortisol stimulates surfactant synthesis by the type II cells of the fetal lung, allowing epinephrine to release the surfactant. In addition, cortisol stimulates production of liver glycogen in the last 15 days of pregnancy. Glycogen reserves contribute to the maintenance of fetal body temperature in the first few days following birth. Cortisol also plays a role in controlling fetal erythropoiesis. In the sheep, maturation of the HPA axis is essential for labor onset. Many other stimulatory factors, however, augment the myometrial contractile activity characteristic of labor. Parturition is not the result of a single signal. Rather, it is the result of a progressive maturation of multiple pathways that may culminate to coordinate and synchronize fetal maturation, myometrial contractile activity, and preparation of the birth canal.

I. INTRODUCTION

A. Pregnancy and Parturition Review

The successful outcome of labor and delivery is dependent upon three processes: maturation of fetal physiological support systems to a level consistent with survival *extra utero,* remodeling of the extracellular matrix of the cervix and uterus, and development of appropriate uterine contractions. In all mammalian species studied to date, these three prerequisites of successful labor and delivery must be satisfied or subsequent neonatal outcome may be compromised. The mechanisms by which these requisite criteria are satisfied and the extent to which they are manifested, however, varies between individual species and is appropriate for the intensity of postnatal care provided (e.g., whether the newborn is altricial or precocious in its development).

With respect to pregnancy and parturition models, the sheep represents one of the most intensely studied species, and one of the few species in which the events culminating in the initiation of labor are relatively well-characterized. The purpose of this chapter is to review recent advances in our understanding of the mechanisms involved in ovine parturition. In particular, the regulation of the fetal HPA axis and the regulation of placental and fetal membrane prostaglandin-forming enzymes will be discussed. The reader is also referred to other excellent recent reviews focusing on ovine parturition (Challis and Lye, 1986; Challis and Olson, 1988; Liggins and Thorburn, 1994; Thorburn and Liggins, 1994).

B. Ovine Parturition: The Current Paradigm

In sheep, pregnancy represents an extension of the luteal phase of the estrous cycle induced by products of the conceptus. These products control the activity of the myometrium during pregnancy, maintaining a quiescent state that is essential for normal fetal development. As term approaches, electromyographic activity increases and synchronizes to produce effective uterine contractions at the time of parturition. Increased uterine activity at term is associated with the induction of placental 17α-hydroxylase/17,20 lyase cytochrome P_{450} (P_{450} 17α), that catalyzes the conversion of C_{21} (progestogenic) precursors to C_{19} (androgenic) steroids. C_{19} steroids are subsequently converted to estrogens (the final synthetic product of the steroid pathway) by placental aromatase. As aromatase activity is not considered rate-limiting in the formation of estrogens, the regulation of placental P_{450} 17α activity is a central determinant. Consistent with this role, the induction of placental P_{450} 17α activity is associated with a dramatic decline in maternal circulating concentrations of progesterone (Figure 1).

In sheep, it is clearly established that products of the fetal adrenal regulate the expression of placental P_{450} 17α. The participation of fetal glucocorticoid in the initiation of ovine parturition was first recognized by the association of prolonged gestation with the congenital absence of the fetal adrenal gland. Subsequent studies have confirmed the involvement of adrenal glucocorticoid synthesis and secretion in the induction of parturition. In addition, fetal glucocorticoid has been implicated in the maturation of fetal organ systems necessary for survival *extra utero*. Thus, the timing of labor onset is primarily dependent upon maturation of the fetal HPA axis. This process can be accelerated by administration of ACTH (Liggins, 1968 cited in Thorburn and Liggins, 1994) or CRH (Brooks et al., 1986 cited in Liggins and Thorburn, 1994) to the fetus. Secretion of CRH and AVP from the fetal hypothalamus stimulates ACTH secretion from the fetal pituitary gland in the last 20 days of pregnancy, although the magnitude of the ACTH response to exogenous CRH and AVP diminishes toward term (Norman and Challis, 1987). During normal pregnancy, fetal plasma cortisol concentrations begin to rise from day 125 of gestation. There is now evidence that this rise is preceded by and is associated with a concurrent increase in the concentration of immunoreactive (ir-) ACTH (Hennessey et al., 1982 cited in Challis and Lye, 1986), although this has not been a consistent finding. The heterogeneity of ACTH assays has hindered the clarification of the

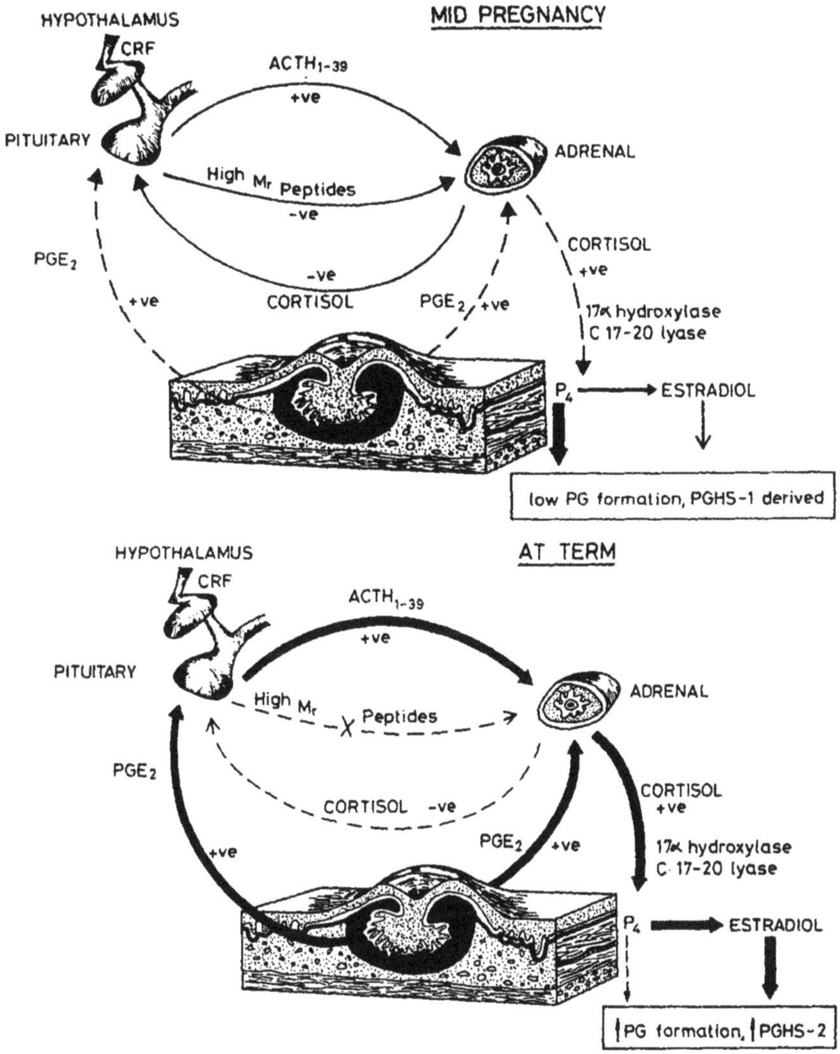

Figure 1. A model depicting the interrelationships between the fetal HPA axis and the placenta at mid-pregnancy (Panel A) and at term (Panel B). During mid-pregnancy, fetal adrenal activity is regulated by a balance between stimulatory ($ACTH_{1-39}$) and inhibitory (High M_r Peptides) factors released from the anterior pituitary. Cortisol released from the adrenal participates in a negative feedback pathway to limit the release of these peptides. Term is characterized by increased release of $ACTH_{1-39}$ and adrenal activity. Concomitant increased cortisol release and a reduced negative feedback effect on the fetal pituitary results in a positive feed-forward cascade. Fetal cortisol induces 17α-OH/C 17-20 lyase activity in the placenta, resulting in the conversion of C_{21} steroids (progestagens) to C_{19} steroids (androgens). Androgens are then readily converted to C_{18} estrogenic steroids. Associated with this change in steroid metabolism is the induction of PGHS-2 and increased eicosanoid formation. (+ve:stimulation; –ve: inhibition.)

temporal relationship between the gestation-related increases in ACTH and cortisol. The fetus, via hypothalamic-pituitary activation alters steroid synthetic pathways in both the fetal adrenal and placenta. As a consequence, progesterone concentrations fall and estrogen synthesis increases. Estrogen has a wide range of physiological functions, including stimulation of prostaglandin production in the placenta and fetal membranes. Myometrial responsiveness to agonists such as oxytocin and prostaglandins is enhanced through increased receptor expression. Thus, there is a progressive increase of myometrial contractility, resulting in expulsion of the fetus.

II. THE FETAL HYPOTHALAMIC-PITUITARY-ADRENAL AXIS

A. Overview

The role of the ovine HPA axis in the timing of labor onset first became evident in 1963 with the recognition that pregnancy was prolonged following maternal ingestion of the teratogenic plant, or skunk cabbage (*Veratrum californicum*), on the fourteenth day of gestation. The toxic, principle was identified as 11-deoxyjervine, a steroidal alkaloid. The affected lambs had pituitary glands, however, the neural connections were abnormal or completely absent (Kennedy, 1971 cited in Thorburn and Liggins, 1994). Although this data provided evidence for the involvement of the fetus in the aetiology of delayed parturition, the precise mechanism responsible for this occurrence still remained to be established.

Professor G.C. Liggins was the first investigator to precisely define the contribution of the fetal pituitary gland to labor. It was demonstrated that electrocoagulation of the fetal pituitary prolonged pregnancy indefinitely and was associated with hypoplasia of the adrenal cortices. Bilateral adrenalectomy of the fetus also prolongs labor onset. Conversely, infusion of glucocorticoid into the fetal lamb *in utero* leads to premature delivery (refer to Thorburn and Liggins, 1994).

In normal pregnancy, the concentration of cortisol in fetal plasma rises from day 125 of gestation. Prior to day 120 of gestation, cortisol in fetal plasma is primarily derived by transplacental transfer from the mother (Hennessey et al., 1982 cited in Challis and Lye, 1986). Fetal glucocorticoids play an important role in lung and organ maturation. Challis and Hooper (1990) proposed that cortisol may enhance placental production of CRH, PGE_2, and ACTH, providing additional trophic support to the fetal pituitary and adrenal glands. Paradoxically, cortisol acts to promote secretion of corticosteroid-binding globulin (CBG) in the fetal circulation (Fairclough and Liggins, 1975 cited in Challis and Lye, 1986). It has been proposed that this may reduce the efficacy of negative feedback at the level of the fetal hypothalamus and pituitary.

B. Fetal Pituitary Function

Previous investigations have demonstrated that the late gestational rise in fetal ACTH concentration provides trophic support for corticosteroidogenesis. Fetal hypophysectomy leads to hypoplasia of the adrenal gland and delayed labor onset. Conversely, the administration of bioactive ACTH in high concentrations stimulates cortisol release from the fetal adrenal gland.

Using immunohistochemistry, the fetal *pars distalis* has been shown to contain ACTH staining cells by day 40 of gestation (Mulvogue et al., 1986 cited in Thorburn and Liggins, 1994). Anterior pituitary cells were shown to contain fragments of pro-opiomelanocortin (POMC), including pro-γ-MSH (melanocyte stimulating hormone), γ-MSH, and β-endorphin by day 38 of gestation. By day 90, the corticotrophs fall into two morphologically distinct cell types: "adult" and "fetal" corticotrophs. Fetal corticotrophs are characteristically large columnar cells which stain with variable intensity for ACTH. The adult corticotroph exhibits intense staining for ACTH and is stellate in shape. Cortisol influences the proportion of adult:fetal corticotrophs, which increases with gestational age. Intrafetal cortisol infusion accelerates the production of adult corticotrophs in the *pars distalis*. In contrast, fetal adrenalectomy delays this change in the ratio of "adult" to "fetal" cell types (McMillen et al., 1990). This may constitute a positive feedback loop between the fetal pituitary and adrenal glands (Figure 2).

The fetal pituitary responds to CRH and AVP by releasing ACTH. AVP acts synergistically with CRH to provoke ACTH release and is more effective earlier in gestation than later. As term approaches, CRH is more effective in inducing ACTH release, as a result of increased CRH receptor expression. At this time, the previous synergism between CRH and AVP is lost (Norman and Challis, 1987). The diminution of the ACTH response following administration of CRH and AVP as term approaches may be explained by the corresponding increase in the cortisol response over the same period. As term approaches, the cortisol concentration

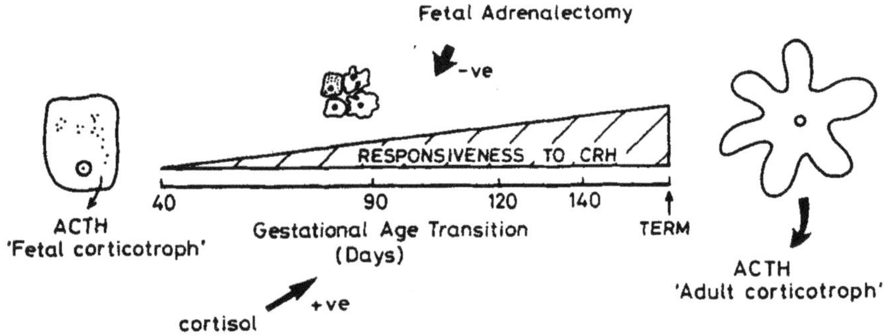

Figure 2. Corticotroph gestational age transition. During gestation, the phenotype of corticotrophs changes dramatically. Fetal corticotrophs are characteristically large columnar cells that stain with variable intensity for ACTH. By 90 days of gestation, two morphologically distinct cell types are identifiable: "adult" and "fetal." The adult phenotype stains intensely for ACTH and displays increased responsiveness to CRH stimulation.

increases more rapidly and to a greater extent following secretagogue administration and this probably terminates ACTH secretion by negative feedback (Figure 1).

POMC undergoes extensive post-translational processing to yield a variety of peptides including $ACTH_{1\text{-}39}$, γ-MSH, and β-endorphin. McMillen and coworkers 1988 (cited in McMillen et al., 1990) measured POMC mRNA in the fetal sheep anterior pituitary gland during fetal development. Despite elevated plasma $ACTH_{1\text{-}39}$ concentrations in late gestation, they found a substantial decline in POMC mRNA at day 142 of gestation compared to day 133. This decline in POMC mRNA appears to result from high circulating concentrations of glucocorticoids present during late pregnancy. Evidence supporting this proposal was obtained from a study in which adrenalectomy at day 120 of gestation resulted in high amounts of POMC mRNA in anterior pituitary gland 15 days later (McMillen et al., 1990). If POMC gene expression increases toward term, the magnitude of the claimed increase is not large in relation to that of fetal cortisol over the same period. This suggests the existence of other factors which determine adrenal activity in the last month of gestation. Increased post-translational processing of POMC may be such a factor (Saphier et al., 1993). In this respect, the expression and activities of the enzymes which process POMC to smaller, bioactive forms may be more significant than the expression of POMC itself. These enzymes exhibit developmental changes in the fetal rat suggesting that similar changes may occur in other species.

Parturition in sheep is preceded by activation of the fetal adrenal gland. ACTH binding sites on the fetal adrenal gland increase markedly from 140 days to term. There is a close temporal relationship between fetal plasma $ACTH_{1\text{-}39}$ concentrations and ACTH receptor concentration. Recent studies suggest that ACTH acts through two distinct receptor mechanisms (Li et al., 1989). One class of receptor is linked to an adenylate cyclase signal transduction pathway. The other class of receptor is linked to an inositol phosphate (IP) and Ca^{2+} stimulated pathway. Li and coworkers (1989) demonstrated that $ACTH_{1\text{-}10}$ binds to the adenylate-cyclase linked receptor but not the receptor linked to IP. Conversely, $ACTH_{11\text{-}24}$ was shown to preferentially bind to the IP-linked receptor.

Evidence suggests that IP-linked receptors are predominant in fetal membranes before day 130 of gestation. Durand found that in fetal adrenal membranes obtained from sheep between 110 and 130 days of gestation, $[^{125}I]$-$ACTH_{1\text{-}24}$ was displaced by $ACTH_{11\text{-}24}$ but not by $ACTH_{1\text{-}10}$ (see Challis and Olson, 1988). This finding is in agreement with adenylate cyclase activity being low in fetal adrenal membranes before day 140 of gestation. Prior to this time, $ACTH_{1\text{-}24}$ binding to adrenal membranes remains low. The rise in ACTH receptor numbers in the sheep fetus 5–7 days before delivery is associated with an increase in cAMP concentration (Durand et al., 1981 as cited in Challis and Olson, 1988). This most likely reflects an increase in the number of adenylate cyclase-linked receptors. Thus, parturition in the ewe is preceded by an increase in the number of low-affinity, strongly steroidogenic receptors (cAMP-associated). These receptors replace the function of the high-affinity, weakly steroidogenic receptors (IP-associated), culminating in increased fetal adrenal cortisol output capable of causing labor onset.

Jones and coworkers proposed a mechanism for the rise in adrenal cortisol output late in gestation (as cited in Challis and Lye, 1986). These investigators demonstrated that three large molecular weight peptides (50, 30, and 20 kDa) present in fetal pituitary late in gestation antagonized the stimulatory action of $ACTH_{1\text{-}39}$ on dispersed sheep adrenal cells. Since the concentration of these large molecular weight peptides relative to $ACTH_{1\text{-}39}$ declines late in pregnancy, it was suggested that the rise in fetal cortisol output was the result, in part, of the withdrawal of the effects of these inhibitory peptides. In support of this hypothesis, recent unpublished work in this laboratory has demonstrated the presence of a pituitary-dependent inhibitor of adrenal function late in gestation. These investigators examined the secretion of cortisol from the fetal adrenal in response to increasing concentrations of $ACTH_{1\text{-}24}$ following hypophysectomy. To prevent adrenal gland atrophy following hypophysectomy, fetal sheep were given a continuous maintenance infusion of $ACTH_{1\text{-}24}$ (34.4 pmol/h/kg). Jacobs et al. (1994) demonstrated that hypophysectomized fetuses receiving this low dose ACTH infusion have maintained adrenal growth and deliver at normal term. In intact fetuses, plasma concentrations of cortisol increased in a dose-dependent manner in response to increasing concentrations of $ACTH_{1\text{-}24}$ at days 130 and 140 of gestation. Similarly, hypophysectomized fetuses responded to exogenous $ACTH_{1\text{-}24}$ administration with increased plasma concentrations of cortisol, however, maximal concentrations achieved were significantly higher than those observed in intact animals. As in the study of Jacobs et al. (1994), the hypophysectomized fetuses receiving chronic, low dose ACTH infusion went on to deliver at term. It was concluded that the removal of the fetal pituitary gland releases the fetal adrenal from endogenous inhibition, thus allowing greater stimulation of cortisol by ACTH. The nature of the putative inhibitory factors *in vivo* is yet to be elucidated.

C. Fetal Adrenal Function

Several investigators have studied the changes in ACTH sensitivity of the adrenal gland in late gestation in the ovine fetus (Challis and Olson, 1988). Three phases of responsiveness to trophic hormonal stimulation have been described. At days 50–60 of pregnancy, the fetal adrenal secretes large amounts of cortisol in response to ACTH and dibutyryl cyclic AMP. This responsiveness is lost in mid-gestation (day 100), but reemerges near term. At term, there is an increased number of ACTH receptors, an increased activity of adenylate cyclase, and enhanced coupling by Gs proteins (Saez et al., 1984 cited in Challis and Hooper, 1990). At day 100 of gestation fetal adrenal cells have the ability to secrete progesterone but not cortisol in response to cAMP. This implies the existence of an enzyme lesion between these two steroids. In early pregnancy, pregnenolone, 17α-hydroxypregnenolone, progesterone, and 17α-hydroxyprogesterone are all converted to cortisol. In mid-pregnancy, only 17α-hydroxyprogesterone is converted to cortisol. By term, the conversion of 17α-hydroxyprogesterone and progesterone to cortisol is increased substantially. These results show that 17α-hydroxylase is a major rate-limiting enzyme for cortisol production and that its activity is increased substantially toward term.

The reason for the recrudescence of adrenal responsiveness around day 120 of gestation is unclear, but the phenomenon is probably fundamental to the mechanism of parturition. Between 90 and 120 days of gestation, placental production of progesterone, estrogens, and PGE_2 increase markedly. This increase in placental hormone production depends on fetal pituitary function (Deayton et al., 1993). PGE_2 is a potent ACTH secretagogue whose activity does not diminish as adrenal maturation proceeds. The tonic increase in PGE_2 secretion which immediately precedes the activation of the fetal HPA axis may thus be a factor in the maturation of the fetal pituitary. PGE_2 has also been reported to directly stimulate cortisol secretion by the fetal adrenal glands (Thorburn and Liggins, 1994). Thus, a positive feedback loop between the fetal pituitary, adrenal gland, and the placenta may be proposed.

Evidence has been obtained that suggests cortisol may be involved in the positive feed-forward glucocorticoid cascade, acting to modulate the mechanism by which ACTH stimulates adrenal activity. Liggins and coworkers first demonstrated in 1977 that the output of cortisol *in vivo* in response to an ACTH challenge was greater in fetuses that had been treated for 48 hours with dexamethasone (cited in Thorburn and Liggins, 1994). Challis and coworkers showed that adrenal cells prepared from fetuses pretreated *in vivo* with ACTH and metopirone (an inhibitor of 11β-hydroxylase and hence cortisol production) had an attenuated capacity to produce cAMP or generate 11-desoxy-cortisol compared to adrenal cells from fetuses treated with ACTH alone. These inhibitory effects on cAMP and steroid production could be overcome by small amounts of replacement cortisol. Larger amounts of cortisol or dexamethasone were found to be less effective or ineffective in reversing the metopirone block (cited in Challis and Hooper, 1990). Thus, small amounts of cortisol appear to modulate the effect of ACTH in activating fetal adrenal function. At higher concentrations cortisol has an inhibitory effect.

The preparturient increase in fetal plasma glucocorticoids first described by Bassett and Thorburn (1969) has been reconfirmed many times. Cortisol concentrations begin to rise 25 days before delivery (Challis and Brooks, 1989 cited in Challis and Hooper, 1990), coinciding with an enhanced responsiveness of the fetal adrenal to $ACTH_{1\text{-}24}$. In accordance with the findings presented above, Jones et al. (1977) showed that although hypoxemia raised the ACTH concentration in fetal plasma, only late in gestation was the increase in ACTH associated with a large and rapid rise in plasma corticosteroid concentrations. Similarly, hemorrhage and hypotension raise plasma concentrations of ACTH, however, there is no significant corticosteroid response until late in gestation (Rose et al., 1978 and 1981 as cited by Challis and Olson, 1988). ACTH stimulates steroidogenesis in the fetal adrenal by at least two different mechanisms. Both involve specific cell membrane receptors which act in a different manner. One class of receptor mobilizes intracellular calcium while the other stimulates the formation of cAMP.

During hypoxemia, there is a 2–3-fold increase in blood flow to the fetal adrenal gland, which is temporally related to a rise in fetal plasma ACTH concentration (Challis and Lye, 1986). Even when ACTH concentrations decline the adrenal blood flow remains elevated. Using immunohistochemistry it was found that the endothelial cells of the subcapsular blood vessels contained the prostaglandin-forming enzyme prostaglandin G/H synthase. Thus, locally stimulated PG production may cause changes in blood flow. Other vasoactive peptides, including CRH and endothelin may be produced in the fetal adrenal to act as effectors of blood flow adaptations.

There is both *in vivo* and *in vitro* evidence that fetal adrenal maturation can be provoked by the administration of $ACTH_{1\text{-}24}$. The activation of adenylate cyclase in fetal adrenal cell membrane preparations is enhanced following ACTH perfusion in fetal sheep. Durand et al. showed in 1982 that ACTH can increase the activity of 11β-hydroxylase and 21-hydroxylase enzymes (cited in Challis and Lye, 1986). Moreover, it was later demonstrated that an extract of fetal and newborn pituitary enhanced the steroidogenic enzyme activity and adenylate cyclase activity by fetal adrenal cells (Durand et al., 1985 cited in Challis and Olson, 1988). The response of adrenals from hypophysectomized fetuses, however, was lower than that of tissue from intact fetuses. It was concluded that the fetal pituitary contained activities that promoted fetal adrenal maturation. This finding appears to contradict the earlier proposal of the large M_r pituitary peptides inhibiting adrenal cortisol output (see Section II B), although there may be a requirement for a small, permissive amount of ACTH or other POMC-derived peptides to prevent adrenal atrophy. $ACTH_{1\text{-}24}$ at a very low concentration is sufficient to maintain fetal growth *in vivo* (Jacobs et al., 1994). Significantly, Durand and coworkers have also demonstrated the spontaneous maturation of fetal adrenal cells in culture, further supporting the proposal that there is an inhibitory factor *in vivo*.

D. The Physiological Role of Cortisol in the Maturation of Fetal Organ Systems

If a lamb is to make the successful transition from intra- to extra-uterine life, the fetal cardiovascular, respiratory, gastrointestinal, and neuroendocrine systems must all undergo adequate maturation *in utero*. The prepartum rise in plasma cortisol concentrations coordinates and induces many of these maturational changes (Thorburn and Rice, 1987).

The ability of a newborn animal to sustain adequate respiratory function is of primary importance to its survival. The production of lung surfactant prior to delivery is thus essential. In fetal sheep, production of pulmonary surfactant occurs from day 120 of gestation and increases to term. This can be correlated temporally with the rise in fetal plasma concentrations of cortisol. Evidence supporting the importance of cortisol to lung maturation was provided by Professor G.C. Liggins in 1969 (cited in Thorburn and Liggins, 1994). When parturition was induced prematurely by the infusion of glucocorticoids at day 135 of gestation, lambs developed a normal respiratory pattern postpartum. In contrast, lambs delivered by cesarean section prior to day 140 of gestation developed severe respiratory distress and were unable to adequately ventilate their lungs. These data indicate that glucocorticoids stimulate production of pulmonary surfactant. This ensures adequate maturation of the fetal lungs to sustain life following birth.

There is evidence to suggest that cortisol plays a role in controlling fetal erythropoiesis. Throughout gestation, hemoglobins synthesized by fetal red blood cells undergo maturational changes (Wood et al., 1976). Two embryonic hemoglobins synthesized by the early conceptus are replaced in the initial stages of pregnancy by fetal hemoglobin. From day 125, the γ chains of fetal hemoglobin are progressively replaced by β chains which characterize adult hemoglobin. The change from fetal to adult hemoglobin occurs concurrently with the loss of the liver as an erythropoietic organ. Wood et al. (1976) suggested that cortisol may initiate β chain synthesis by fetal reticulocytes as the prepartum rise in cortisol was temporally related to the structural change. The infusion of dexamethasone into the fetus causes a rapid switch in hemoglobin synthesis after day 130 of gestation. Fetal hypophysectomy, on the other hand, slows the rate of switching from fetal to adult forms (Wood et al., 1976).

Birth constitutes an abrupt transition for the newborn from the thermal environment of the uterus to the cooler extra-uterine environment. This poses problems for the neonate in the maintenance of its body temperature. Liver glycogen accumulates in the fetal lamb in the last 15 days of pregnancy. This contributes to thermogenesis. The rise in glycogen accumulation in the liver is directly related to the prepartum increase in cortisol production. Cortisol infusion into the intact fetus stimulates production of liver glycogen. In contrast, fetal hypophysectomy or fetal adrenalectomy inhibits accumulation of glycogen. Although brown fat and glycogen contribute to the maintenance of neonatal body temperature, increased production of triiodothyronine (T_3) is required to mobilize these reserves. During fetal life, the concentrations of thyroxine (T_4) and "reverse T_3" (RT_3; the inactive biological isomer of T_3) are high, whereas the concentration of T_3 is low. The increase in fetal concentrations of T_3 and the decrease in fetal T_4 and RT_3 concentrations in late gestation can be temporally correlated to the prepartum rise in fetal cortisol production.

It can be clearly seen that the increase in cortisol production late in gestation is not only essential for birth but also prepares fetal lambs for survival *extra utero*. High circulating concentrations of cortisol ensure that the development of the fetal lamb is sufficient to maintain physiological homeostasis at birth.

III. PLACENTA

A. Progesterone

Sheep require luteal progesterone for the establishment and maintenance of pregnancy (corpus luteum-dependent). The requirement for luteal progesterone exceeds the length of the luteal phase of the estrous cycle, necessitating an extension of the lifespan of the corpus luteum (CL). By day 50 of pregnancy, the placenta secretes sufficient progesterone to support the pregnancy. After this time, ovariectomy can be performed without inducing abortion (CL-independent). A further increase in placental progesterone secretion occurs between days 90 and 120 of gestation, a time when placental growth has ceased but the fetus undergoes rapid growth.

Progesterone concentrations in fetal and maternal plasma decline prior to the onset of labor. Nathanielsz et al. (1982) reported that the mean time of onset of the fall in maternal progesterone was 3.5 ± 0.5 days before parturition. This is coincident with the preparturient surge in fetal plasma cortisol (Thorburn and Liggins, 1994). Fetal plasma concentrations of progesterone closely parallel those of the mother, however, absolute concentrations are generally lower. This can be attributed to vascular metabolism by the 20α-hydroxysteroid dehydrogenase (20α-HSD) enzyme present in fetal erythrocytes.

The importance of progesterone withdrawal as a prerequisite to the onset of labor in sheep has been the subject of many investigations. Progesterone has been shown to inhibit labor at term in sheep (Liggins et al., 1973 cited in Liggins and Thorburn, 1994). Liggins and coworkers showed that progesterone can prevent dexamethasone-induced premature labor when administered in pharmacological but not physiological concentrations. Doses of progesterone below 100 mg/24 h do not significantly delay labor onset in sheep given an intrafetal infusion of dexamethasone, although the dose is sufficient to maintain the peripheral concentration of progesterone within the preparturient range. A dose of 200 mg/24 h is sufficient to block cervical dilatation, uterine activity, and labor. This amount of progesterone, however, is more than twice the maximal production rate for late ovine pregnancy. Recent studies have provided a possible explanation for why such high concentrations of exogenous progesterone are required to block labor. Administration of synthetic glucocorticoids or cortisol is associated with the induction of placental 17α-hydroxylase enzyme. Endogenous pregnenolone metabolism is thus directed away from progesterone formation toward estrogen synthesis. High concentrations of exogenous progesterone are therefore required to counteract the high plasma concentrations of estrogen. In addition, exogenous progesterone represents the sole source of progesterone in the maternal circulation.

Enhanced uterine prostaglandin release can be induced by progesterone withdrawal following the administration of a 3β-HSD inhibitor (Taylor et al., 1982 cited in Challis and Lye, 1986). A decrease in plasma progesterone for six hours was shown to be sufficient to increase circulating concentrations of $PGF_{2\alpha}$ as indicated by 13,14-dihydro 15-keto $PGF_{2\alpha}$ production (PGFM, the inactive metabolite of $PGF_{2\alpha}$) and premature delivery. The importance of progesterone withdrawal as a prerequisite to the onset of parturition was highlighted in a study conducted by Thorburn and coworkers (1984 cited in Liggins and Thorburn, 1994). $PGF_{2\alpha}$ was infused into the ewes via an extra-amniotic catheter at concentrations sufficient to raise maternal, fetal, and amniotic fluid PGFM concentrations to those observed at term. This did not induce a sustained increase in uterine activity unless progesterone concentrations were decreased by the administration of a 3β-HSD inhibitor. Lye and Porter (1978) observed that progesterone could inhibit the myometrial activity induced by the administration of $PGF_{2\alpha}$ to estrogen-treated, ovariectomized ewes.

B. Estrogen

A role for estrogen in the initiation of labor in sheep has largely been supported by studies employing two lines of investigation, namely those based on measurement of circulating concentrations of estradiol and those obtained by the administration of exogenous estrogen to late pregnant ewes. Maternal peripheral total unconjugated estrogen concentrations (i.e., estradiol-17α, estradiol-17β, and estrone) remain low throughout pregnancy and rise 24 hours before delivery (Challis and Olson, 1988). Unconjugated estrogens also rise in amniotic fluid during the last 3–6 days of pregnancy. A close temporal relationship between the rising concentrations of estradiol-17β (E_2-17β) and $PGF_{2\alpha}$ was described by Currie et al. (1973 cited in Challis and Olson, 1988). In addition, Liggins et al. (1973 cited in Liggins and Thorburn, 1994) demonstrated that the concentration of $PGF_{2\alpha}$ in utero-ovarian vein plasma rises parallel with estrogen output following spontaneous labor onset and also following premature delivery induced by dexamethasone or ACTH.

Hindson et al. (1967) were the first investigators to demonstrate labor induction in sheep following exogenous estrogen administration. Sheep given 20 mg of stilboestrol during mid- or late-pregnancy gave birth prematurely. This estrogen-induced labor, however, did not result in "normal" parturition. Delivery was frequently delayed due to failure of the cervix to dilate (Hindson et al., 1967). Liggins suggested that the observed failure of cervical softening may have been associated with the fact that plasma progesterone concentrations did not fall following estrogen administration. The finding that 150 mg/day of progesterone given to ewes is not sufficient to block dexamethasone induced labor but is associated with failure of cervical dilation supports this view (Liggins et al., 1972 as cited in Liggins and Thorburn, 1994).

In vitro and *in vivo* evidence has shown that fetal cortisol induces placental 17α-hydroxylase and $C_{17\text{-}20}$ lyase enzyme activity. Thus, the concentration of androstenedione and of dehydroepiandrosterone (DHA) increases in maternal utero-ovarian vein plasma coincident with the preparturient rise in estrogen output. There is a correlation between the rise in maternal unconjugated estrogen concentration and the increase in estrogen concentration in the endometrium and myometrium. Liggins et al. (1973, cited in Liggins and Thorburn, 1994) reported a rise in the concentration of $PGF_{2\alpha}$ in uterine venous plasma and in myometrium following estrogen-induced labor. Despite the absence of any change in the concentration of progesterone in both peripheral and utero-ovarian venous plasma, the observed rise in $PGF_{2\alpha}$ was associated with a 90% reduction in the threshold of the excitatory myometrial response to oxytocin. This experimental evidence implies that the effect of estrogen on the myometrium is mediated by an increase in uterine prostaglandin biosynthesis at term in sheep. The evidence, however, is conflicting. Jenkin and Thorburn (1985) demonstrated labor induction in sheep following the administration of the 3β-HSD inhibitor, trilostane. There was a major reduction in circulating concentrations of progesterone and enhanced $PGF_{2\alpha}$ synthesis without an alteration in estrogen concentrations. Furthermore, labor induction by intra-fetal administration of glucocorticoid can occur despite no elevation in maternal concentrations of estrogens (Kendall et al., 1977 cited in Challis and Lye, 1986).

IV. OXYTOCIN

Oxytocin may also play a role in successful parturition in the ewe. Oxytocin is one of the most potent natural substances known to stimulate uterine contractions. Although oxytocin receptor concentrations in uterine endometrium increase in late gestation, recent data suggests that oxytocin exerts its stimulatory effect through an action on uterine prostaglandins.

The release of $PGF_{2\alpha}$ in response to oxytocin increases progressively throughout gestation and reaches maximum concentrations at the time of parturition. It has been suggested that this elevation in prostanoid production is due to a decrease in the inhibitory effect of progesterone on oxytocin receptors or on the transduction of the oxytocin signal to initiate $PGF_{2\alpha}$ release. Treatment of late pregnant ewes with the progesterone receptor antagonist RU486 causes an increase in basal concentrations of both PGE_2 and PGFM. In addition, the subsequent administration of oxytocin induces the release of uterine PGFM, but not PGE_2 (Burgess and Jenkin, 1992 cited in Liggins and Thorburn, 1994). The lack of effect of oxytocin on PGE_2 concentrations indicates a differential effect of oxytocin on prostaglandin synthesis. Oxytocin mediates its effects by interacting with its receptor. Fetal cotyledons, amnion, chorion, myometrium, and cervix have all been shown to produce $PGF_{2\alpha}$ and PGE_2 (Evans et al., 1982 cited in Liggins and Thorburn, 1994). Thus, not all uterine tissues responsible for PGE_2 release may be able to respond to oxytocin. In view of this finding, it is feasible to suggest that the localization of oxytocin receptors may influence the relative proportion of $PGF_{2\alpha}$ and PGE_2 release in late gestation in response to oxytocin. Certainly, the prevailing endocrine milieu at the time of parturition in the ewe would be favorable for increased synthesis of uterine oxytocin receptors.

To further investigate the relationship between oxytocin and PGs in late pregnancy in sheep, Jenkin examined the effect of the oxytocin receptor antagonist, CAP, on basal and oxytocin-induced fetal and maternal PG release (see Jenkin, 1992). The intra-arterial infusion of CAP at 100 $\mu g\ min^{-1}$ did not suppress fetal or maternal PGE_2 or PGFM concentrations at any stage of gestation. CAP did, however, inhibit oxytocin-induced PGFM release and oxytocin-induced uterine activity in late gestation.

V. PROSTAGLANDINS

The importance of prostaglandins to the onset and maintenance of pregnancy has been unequivocally established. Prostaglandins of uterine origin have a central role in controlling pregnancy in the ewe. For example, increased prostaglandin synthesis during the luteal phase of the estrous cycle causes luteolysis and prostaglandin release. At the time of parturition, prostaglandins stimulate forceful contractions of the uterus.

A. Prostaglandin Biosynthesis

Prostaglandins are synthesized from C20 polyunsaturated fatty acids, including 5,8,11,14-eicosatetraenoic acid (arachidonic acid) which is a member of the n6 family of essential fatty acids. Arachidonic acid is predominantly found in tissues in an esterified form and is released from cellular stores by the action of phospholipases. Arachidonic acid is liberated from phospholipid stores primarily by two pathways involving phospholipase A_2 or phospholipase C. Phospholipase A_2 (PLA_2) directly hydrolyzes arachidonic acid esterified in the sn-2 position of phospholipids. Phospholipase C hydrolyses the polar head group of phospholipids, displaying a substrate preference for phosphotidylinositol. The resultant diacylglycerol (DAG) may be further metabolized by diacylglycerol- and monoacylglycerol-lipases to yield non-esterified fatty acids (including arachidonic acid; Okazaki et al., 1981 cited in Rice,

1995). DAG may also function to activate PLA_2 via the phosphorylation of lipocortins (inhibitors of PLA_2 activity) or mobilization of calcium (Ca^{2+}) in response to 1,4,5-inositol triphosphate (IP_3). The net result of both pathways is an increase in the intracellular availability of arachidonic acid for further processing by enzymes such as prostaglandin G/H synthase or lipoxygenase enzymes.

The conversion of arachidonic acid into prostaglandins is catalyzed by the prostaglandin endoperoxidase G/H synthase (PGHS). PGHS is an integral membrane protein found in greatest abundance in the endoplasmic reticulum but which is also present in the plasma membrane and nuclear envelope of cells (Rollins and Smith, 1980). The native enzyme is a glycoprotein composed of two identical subunits (M_r 65,600 Da, unglycosylated; DeWitt and Smith, 1988) and is a protoporphyrin IX-containing protein (Hemler and Lands, 1976 cited in Rice, 1995) with one molecule of heme per subunit.

PGHS catalyzes the initial steps in the formation of prostanoids, causing the conversion of arachidonic acid into PGH_2. Two distinct catalytic activities are involved. The first is a *bis*-oxygenase activity (cyclooxygenase) which catalyzes the oxygenation of arachidonic acid resulting in the formation of PGG_2. This process involves the incorporation of two molecules of oxygen. A peroxide link is formed between C_9 and C_{11} of arachidonic acid by one molecule of oxygen, while the other is incorporated as a hydroperoxide at C_{15}. The second activity of PGHS is a hydroperoxidase activity that catalyzes the reduction of the 15-hydroperoxyl group of PGG_2 to form PGH_2 (Pagels et al., 1983).

B. Prostaglandin Synthase Isozymes

Recent advances in our understanding of the biochemical pathway involved in the conversion of arachidonic acid into the prostaglandins have provided new insights into regulation of prostaglandin synthesis by gestational tissues. Of particular relevance to the processes involved in labor onset and maintenance has been the identification of multiple forms (isozymes) of prostaglandin G/H synthase. These PGHS enzymes are thought to represent constitutive (i.e., involved in routine cellular metabolism, PGHS-1) and inducible (i.e., agonist-stimulated, PGHS-2) isozymes.

A new epoch in prostaglandin biochemistry began with the cloning of the ram seminal vesicle PGHS (PGHS-1; Merlie et al., 1988) and the subsequent discovery and cloning of the PGHS-2 gene (Kujubu et al.,

1991). PGHS-1 is encoded by a mRNA transcript of ~2.8 kilobases (kb) which represents the PGHS isozyme purified from ram seminal vesicles more than 15 years ago. All investigations of the expression of PGHS in ovine intrauterine tissues before the cloning of PGHS-2 utilized reagents developed against PGHS-1. It is likely that most of the polyclonal antibodies raised against ram seminal vesicle PGHS-1 used in previous studies are incapable of distinguishing between PGHS-1 and PGHS-2. Thus, the data are potentially confounded by the differential regulation and/or expression of PGHS-1 and PGHS-2.

PGHS-2 (70 kDa) is encoded by a mRNA transcript of ~4.0–4.5 kb and displays approximately 64% amino acid sequence homology with PGHS-1 (O'Banion et al., 1992). The important functional and structural domains of PGHS-1 are highly conserved in PGHS-2, including the aspirin acetylation site, the active-site tyrosine, a hydrophobic trans-membrane domain, and the N-linked glycosylation site. Unlike PGHS-1, PGHS-2 is mitogen-induced and is inhibited by glucocorticoid in some tissues (DeWitt and Meade, 1993). It is likely that the regulation of the inducible isozyme is, however, tissue specific. The PGHS-2 isozyme is expressed following cellular activation by stimulatory agonists such as growth factors, phorbol esters, and bacterial exotoxin (O'Sullivan et al., 1992; Hamasaki et al., 1993). In other tissues PGHS-2 has been reported to be induced by pro-inflammatory mediators, such as bacterial endotoxin and cytokines (e.g., TNF and IL-1). Thus, it is most likely that PGHS-2 is induced at the time of bacterial-infection-associated pre-term labor and possibly in association with normal spontaneous onset labor at term.

Recent studies have shown a differential regulation of PGHS-1 and PGHS-2 expression. Wimsatt et al. (1993) reported that the tissue content of immunoreactive PGHS-1 does not change during late gestation or in association with labor in the ewe. This occurs despite increased prostaglandin production at the time of parturition. These data are consistent with the suggestion that PGHS-2 may be the principal isozyme contributing to prostanoid formation at the time of labor. In support of this hypothesis, we have demonstrated induction of PGHS-2 enzyme in ovine cotyledon following intrafetal administration of the glucocorticoid betamethasone.

C. Prostaglandins and Parturition

Normal parturition in the sheep is associated with a marked increase in concentrations of $PGF_{2\alpha}$ in utero-ovarian vein plasma (Mitchell et al.,

1979) and uterine tissues (Risbridger et al., 1985 cited in Challis and Olson, 1988). Mitchell et al. (1979) observed increased concentrations of PGE_2, PGFM, and 6-keto-$PGF_{1\alpha}$ (the major metabolite of PGI_2) before spontaneous labor. Similarly, the concentration of PGE_2 in fetal plasma and of PGE_2, $PGF_{2\alpha}$, and 6-keto-$PGF_{1\alpha}$ in amniotic fluid increases before delivery (Challis and Lye, 1986).

In vitro studies have demonstrated that uterine prostaglandin output remains low until the later stages of gestation. Olson et al. (1986) measured concentrations of PGE_2, $PGF_{2\alpha}$, PGFM, 6-keto $PGF_{1\alpha}$, and 6-keto PGE_1 at day 131 of pregnancy. PGE_2 concentrations were high in chorioallantois and the fetal portion of the cotyledons. PGFM concentrations were high in maternal endometrium. Moreover, Evans et al. (1982 as cited by Liggins and Thorburn, 1994) measured intrauterine tissue concentrations of PGE_2, $PGF_{2\alpha}$, and 6-keto $PGF_{1\alpha}$ at different stages of pregnancy in sheep. No difference in the tissue content of these PGs was observed at days 50, 100, and 130 of gestation. The concentrations of these PGs in cotyledons, chorioallantois, and amnion, however, were all significantly higher on days 130 and 145 than on days 50 or 100 of pregnancy. Risbridger et al. (1985 as cited by Challis and Olson, 1988) examined the *in vitro* capacity of ovine trophoblast cells to synthesize PGs. The synthesis of $PGF_{2\alpha}$, PGE_2, and 6-keto-$PGF_{1\alpha}$ was found to be low during early and mid gestation. The synthesis of PGs increased only after 100 days of gestation. A similar profile was observed by Rice et al. (1988). These investigators demonstrated that the prostaglandin synthesizing capacity of cotyledonary microsomes prepared from ewes 21–145 days of gestation increased only in late gestation. On the basis of these data we proposed that PG production by gestational tissues is regulated throughout pregnancy by the expression of PGHS, which is low during early pregnancy and increased at term (Thorburn and Rice, 1987).

VI. MYOMETRIUM AND THE CERVIX

Throughout pregnancy the myometrium remains relatively quiescent, providing an appropriate environment within the uterus for the developing fetus. At term, rhythmic and sustained contractions of the uterine muscle and dilatation of the cervix combine to expel the fetus when it is mature enough for independent life.

A. Myometrial Structure

The sheep uterus is a bicornate structure, consisting of two separate horns and a common segment. The smooth muscle bundles of the uterus are arranged in two layers, which together form the myometrium. The outer layer muscle bundles are arranged longitudinally and shorten the uterus cephalo-caudally upon contracting. The inner layer muscle bundles are arranged concentrically around the long axis of the uterus. Contraction of this layer constricts the uterine lumen. The uterine smooth muscle cell is a spindle shaped and elongated cell at rest and varies in size from 20–600 μm long by 2–10 μm wide. The dimension of a cell is dependent upon its functional state, with the longest cells being observed in term pregnant animals.

During labor, the myometrium must function optimally and develop well-coordinated synchronous contractions. This requires well developed cell–cell coupling. Gap junctions are thought to be responsible for this coupling. These intermembranous pores provide sites for low-resistance electrical or ionic coupling as well as a pathway for the transport of metabolites directly between cells. Throughout pregnancy, the number of gap junctions is low or minimal. At term, the number increases dramatically. This is thought to be facilitated by the changing endocrine milieu, in particular the rise in intra-uterine concentrations of oxytocin. In addition, the decrease in progesterone levels and rise in estrogen production at term causes an increase in the synthesis of proteins associated with gap junctions. The proteins are inserted into the plasma membranes of cells and aggregate to form membranous pores. PGs modulate the aggregation process. $PGF_{2\alpha}$ and PGE_2 stimulate aggregation while prostacyclin (PGI_2) inhibits gap junction formation.

The organization of the smooth muscle cell is different from that of skeletal muscle. The thick myosin and thin actin filaments of the contractile system occur in long, random bundles throughout the smooth muscle cell and the continuity of these filaments is not interrupted by Z lines. Analogous to the Z lines are intermediate filaments, which form a network with dense protein bodies. The intermediate filaments and dense bodies are not directly involved in the contractile process but are distributed throughout the cytoplasm and on the inner surface of the cell membrane. This provides a framework for actin attachment to support integrated cellular contractions. In skeletal muscle, the contraction is aligned with the axis of the muscle fibers. In smooth muscle, the pulling forces can be exerted in any direction, allowing the uterus to assume any shape capable of accommodating the growing fetus.

The above section has provided a brief account of the gross structure of the myometrium. For information regarding the ultrastructure and biochemistry of myometrial contraction, the reader is referred to several excellent reviews (Huszar and Roberts, 1982; Challis and Olson, 1988).

B. Hormonal Control of Myometrial Contractility

Throughout pregnancy, the uterine muscle is relatively quiescent, however, it is never completely inactive. In the sheep, electromyographic (EMG) activity has been recorded at day 64 of gestation (term approximately 145 days). At this time, the episodes of myometrial contraction are characteristically of long duration (four minutes or more) and low-amplitude (<5 mmHg) and occur approximately once per hour. This is not sufficient to expel the uterine contents.

At term, bursts of EMG activity are of one minute duration or less and cause an increase in intra-uterine pressure of greater than 5 mmHg. Contractions occur at a frequency of 30 events/h (Harding et al., 1982 cited in Challis and Olson, 1988). The ability of the uterine muscle to develop strong contractions at term is the result of many factors including the appearance of receptors for oxytocin and estradiol-17β, enhanced synthesis of stimulatory agonists, reduced production of inhibitory factors, a lower resting membrane potential of cells, and improved conduction between cells due to the formation of gap junctions.

The relative inactivity of uterine muscle throughout pregnancy is largely attributed to progesterone. Progesterone has an antagonistic effect on the estrogenic induction of uterine oxytocin receptors. Exogenous administration of progesterone decreases uterine activity and delays parturition. Moreover, progesterone acts to suppress excitation-contraction coupling as well as the formation of gap junctions. The inhibition of myometrial contraction of uterine muscle *in vitro* by progesterone is subject to a time lag (Lye and Porter, 1978). This is consistent with an action on protein synthesis.

Several other factors are responsible for myometrial contractile refractoriness throughout gestation. PGI_2 is a major prostaglandin produced by the ovine myometrium. Exogenous administration in the sheep abolishes spontaneous electrical and mechanical activity within minutes of administration (Lye and Challis, 1982 cited in Challis and Lye, 1986). Moreover, PGI_2 acts to inhibit gap junction formation. Administration of relaxin also inhibits spontaneous myometrial contractions (Porter et al., 1981). Concentrations of relaxin increase steadily during pregnancy. The precise role of relaxin, however, in species where progesterone is inhibitory is yet to be elucidated.

Despite PGI_2 and relaxin acting to inhibit myometrial activity alone, they do not block the stimulatory response of the myometrium to oxytocin or $PGF_{2\alpha}$. Other peptides that may have a role in maintaining uterine quiescence throughout pregnancy include vasoactive intestinal polypeptide (VIP) and catecholamines. The exact nature of the contribution of these peptides is yet to be established.

C. The Cervix

The cervix is a strong, rigid, and collagenous structure that ensures the products of conception are retained within the uterus. In the sheep, the cervix is composed of three structural layers:

1. An inner mucosa and submucosa;
2. A layer of smooth muscle that is continuous with that of the uterus and vagina; and
3. An intermediate layer consisting of collagen, a limited amount of smooth muscle and a connective tissue matrix.

Remodeling of cervical tissue takes place during pregnancy in preparation for dilatation at birth. Stys et al. (1978) demonstrated a rise in cervical compliance following labor induction with dexamethasone. Cervical softening still occurred, however, when progesterone was administered simultaneously in sufficient doses (200 mg/day) to prevent the prepartum fall in this steroid. Labor onset was also inhibited. Thus, the release from progesterone dominance may not be a prerequisite for cervical ripening.

PGs not only enhance uterine contractility but also contribute to cervical effacement and softening. Mitchell and Flint (1978) administered the PGHS inhibitor meclofenamic acid to late pregnant ewes in which labor had been induced by the intra-fetal infusion of dexamethasone. Despite an alteration in the steroid milieu which normally accompanies parturition, delivery did not take place until the inhibitor had been withdrawn. Upon removal of this inhibitor, coordinated myometrial contractions and cervical dilatation ensued. Cervical softening was blocked by the inhibition of PG synthesis. It has been demonstrated previously that when PGE_2 or $PGF_{2\alpha}$ is given by intra-aortic infusion or delivered locally by an intra-cervical catheter, cervical softening occurs (Thorburn and Liggins, 1994).

The sheep cervix has been shown to contain large amounts of PGs *in vitro* (Ellwood et al., 1980). During late pregnancy (105–135 days of gestation), concentrations of $PGF_{2\alpha}$, PGFM, 6-oxo-$PGF_{1\alpha}$, and thromboxane B_2 are low. Production of PGE_2 predominates at this time. At delivery, the pattern of prostanoid production significantly alters. There is an increase in the synthesis of PGE_2 and 6-oxo-$PGF_{1\alpha}$ when compared to late pregnant tissues. Collagen breakdown may be a part of the mechanism of cervical softening in the ewe. The principal enzyme involved in collagen breakdown is collagenase. It is a highly specific enzyme that cleaves the collagen molecule at a particular site, resulting in denatured products susceptible to attack by non-specific proteases. The changes that occur in the cervix during gestation and parturition include an increase in vascularization and mass, and a decrease in the organization of collagen and dermatan sulphate (Fosang et al., 1984). The tissue of the cervix becomes edematous following labor onset. Fosang and coworkers (1984) demonstrated an eightfold increase in cervical mass throughout the course of gestation. The morphology of the cervix did not differ at 100 days gestational age when compared to non-pregnant ewes. At 140 days of gestation, however, there was evidence of the loss of the structural organization of the cervix. This was indicated by frayed collagen fibers, wider spaces between cells, and infiltration of inflammatory cells, lymphocytes, and some eosinophils. At term, the collagen fibers were shown to be small and randomly organized. Cervical morphology had recovered 18 h postpartum. Hydroxyproline is commonly used as an indicator of collagen content. The concentration of this molecule was shown to be similar in non-pregnant animals and at day 100 of gestation. The concentration, however, decreased progressively to term.

Uterine myometrial contractions do not contribute to the change in cervical compliance associated with delivery. In a study by Ledger et al. (1985), the cervices of ewes were transected to effectively isolate them from mechanical and vascular connections with the myometrium. Subsequently, labor was induced in these ewes by the intra-fetal administration of dexamethasone. The ewes were killed and the cervices collected 48 h after dexamethasone administration. The compliance of each cervix was examined. Transected ewes still demonstrated cervical ripening.

These data suggest that biochemical changes that occur prior to delivery are of greater significance to cervical effacement than muscular activity.

VII. SUMMARY

The aim of this chapter was to provide a summary of the biochemical and physiological changes associated with late pregnancy and parturition in the ewe. The data presented clearly implicate the steroids, estrogen, and progesterone, and other stimulatory agonists including oxytocin, ACTH, and cortisol in the complex chain of hormonal events leading to birth. A critical event in determining the timing of labor onset in sheep is the maturation of the fetal HPA axis and the induction of steroidogenic enzymes within the placenta. Together, these changes work synchronously to culminate in myometrial contractile activity capable of expulsion of the fetus and uterine contents.

REFERENCES

Bassett, J.M., & Thorburn, G.D. (1969). Fetal plasma corticosteroids and the initiation of parturition in sheep. J. Endocrinol. 44, 285–286.

Challis, J.R., & Lye, S.J. (1986). Parturition. Oxf. Rev. Reprod. Biol. 8, 61–129.

Challis, J.R.G., & Hooper, S.B. (1990). Birth: Outcome of a positive cascade. In: *Perinatal Endocrinology, Baillere's Clinical Endocrinology and Metabolism* (Jones, C.T., Ed.) pp. 781–793. Saunders, New York.

Challis, J.R.G., & Olson, D.M. (1988). Parturition. In: *The Physiology of Reproduction* (Knobil, E., & Neill, J., Eds.) pp. 2177–2216. Raven Press, New York.

Deayton, J.M., Young, I.R., & Thorburn, G.D. (1993). Early hypophysectomy of sheep fetuses: Effects on growth, placental steroidogenesis and prostaglandin production. J. Reprod. Fertil. 97, 513–520.

DeWitt, D.L., & Smith, W.L. (1988). Primary structure of prostaglandin G/H synthase from sheep vesicular gland determined from the complementary DNA sequence. Proc. Natl. Acad. Sci. USA 85, 1412–1416.

DeWitt, D.L., & Meade, E.A. (1993). Serum and glucocorticoid regulation of gene transcription and expression of the prostaglandin H synthase-1 and prostaglandin H-synthase-2 isozymes. Arch. Biochem. Biophys. 306, 94–102.

Ellwood, D.A., Mitchell, M.D., Anderson, A.B., & Turnbull, A.C. (1980). Specific changes in the *in vitro* production of prostanoids by the ovine cervix at parturition. Prostaglandins 19, 479–488.

Fosang, A.J., Handley, C.J., Santer, V., Lowther, D.A., & Thorburn, G.D. (1984). Pregnancy-related changes in the connective tissue of the ovine cervix. Biol. Reprod. 30, 1223–1235.

Hamasaki, Y., Kitzler, J., Hardman, R., Nettesheim, P., & Eling, T.E. (1993). Phorbol ester and epidermal growth factor enhance the expression of two inducible prostaglandin H synthase genes in rat tracheal epithelial cells. Arch. Biochem. Biophys. 304, 226–234.

Hindson, J.C., Schofield, B.M., & Turner, C.B. (1967). The effect of a single dose of stilboestrol on cervical dilation in pregnant sheep. Res. Vet. Sci. 8, 353–360.

Huszar, G., & Roberts, J.M. (1982). Biochemistry and pharmacology of the myometrium and labor: Regulation at the cellular and molecular levels. Am. J. Obstet. Gynecol. 142, 225–237.

Jacobs, R.A., Young, I.R., Hollingworth, S.A., & Thorburn, G.D. (1994). Chronic administration of low doses of adrenocorticotropin to hypophysectomized fetal sheep leads to normal term labor. Endocrinology 134, 1389–1394.

Jenkin, G. (1992). Oxytocin and prostaglandin interactions in pregnancy and parturition. J. Reprod. Fert. (Suppl.) 45, 97–111.

Jenkin, G., & Thorburn, G.D. (1985). Inhibition of progesterone secretion by a 3 beta-hydroxysteroid dehydrogenase inhibitor in late pregnant sheep. Can. J. Physiol. Pharmacol. 63, 136–142.

Jones, C.T., Boddy, K., Robinson, J.S., & Ratcliffe, J.G. (1977). Developmental changes in the responses of the adrenal glands of the fetal sheep to endogenous adrenocorticotrophin as indicated by the hormone responses to hypoxaemia. J. Endocrinol. 72, 279–292.

Kujubu, D.A., Fletcher, B.S., Varnum, B.C., Lim, R.W., & Herschman, H.R. (1991). TIS10, a phorbol ester tumor promoter-inducible mRNA from swiss 3T3 cells, encodes a novel prostaglandin synthase/cyclooxygenase homologue. J. Biol. Chem. 266, 12866–12872.

Ledger, W.L., Webster, M., Harrison, L.P., Anderson, A.B., & Turnbull, A.C. (1985). Increase in cervical extensibility during labor induced after isolation of the cervix from the uterus in pregnant ewes. Am. J. Obstet. Gynecol. 151, 397–402.

Li, Z.G., Park, P., & LaBella, F.S. (1989). ACTH (1-10) and -(11-24) promote adrenal steroidogenesis by different mechanisms. Endocrinology 125, 592–596.

Liggins, G.C., & Thorburn, G.D. (1994). Initiation of parturition. In: *Marshall's Physiology of Reproduction*, Vol. 4 (Lamming, G.E., Ed.) pp. 863–1002. Chapman and Hall, London.

Lye, S.J., & Porter, D.G. (1978). Demonstration that progesterone "blocks" uterine activity in the ewe *in vivo* by a direct action on the myometrium. J. Reprod. Fertil. 52, 87–94.

McMillen, I.C., Antolovich, G.C., Mercer, J.E., Perry, R.A., & Silver, M. (1990). Proopiomelanocortin messenger RNA levels are increased in the anterior pituitary of the sheep fetus after adrenalectomy in late gestation. Neuroendocrinology 52, 297–302.

Merlie, J.P., Fagan, D., Mudd, J., & Needleman, P. (1988). Isolation and characterization of the complementary DNA for sheep seminal vesicle Prostaglandin Endoperoxide Synthase (Cyclooxygenase). J. Biol. Chem. 263, 3550–3553.

Mitchell, M.D., Anderson, A.B., Brunt, J.D., Clover, L., Ellwood, D.A., Robinson, J.S., & Turnbull, A.C. (1979). Concentrations of 6-oxo-prostaglandin $F_{1\alpha}$ in maternal and fetal plasma of sheep during spontaneous and induced parturition. J. Endocrinol. 83, 141–148.

Mitchell, M.D., & Flint, A.P.F. (1978). Prostaglandin production by intra-uterine tissues from preparturient sheep: Use of a superfusion technique. J. Endocrinol. 76, 111–121.

Nathanielsz, P.W., Elsner, C., Magyar, D., Fridshal, D., Freeman, A., & Buster, J.E. (1982). Time trend analysis of plasma unconjugated and sulfoconjugated estrone and 3β-delta^5-steroids in fetal and maternal sheep plasma in relation to spontaneous parturition at term. Endocrinology 110, 1402–1407.

Norman, L.J., & Challis, J.R.G. (1987). Synergism between systemic corticotropin-releasing factor and arginine vasopressin on adrenocorticotropin release *in vivo* varies as a function of gestational age in the ovine fetus. Endocrinology 120, 1052–1058.

O'Banion, M.K., Winn, V.D., & Young, D.A. (1992). cDNA cloning and functional activity of a glucocorticoid-regulated inflammatory cyclooxygenase. Proc. Natl. Acad. Sci. USA 89, 4888–4892.

O'Sullivan, M.G., Chilton, F.H., Huggins, E.M., Jr., & McCall, C. (1992). Lipopolysaccharide priming of alveolar macrophages for enhanced synthesis of prostanoids involves induction of a novel prostaglandin H synthase. J. Biol. Chem. 267, 14547–14550.

Olson, D.M., Lye, S.J., & Challis, J.R. (1986). Prostaglandin concentrations in ovine maternal and fetal tissues at late gestation. Pediatr. Res. 20, 83–86.

Pagels, W.R., Sachs, R.J., Marnett, L.J., DeWitt, D.L., Day, D.S., & Smith, W.L. (1983). Immunochemical evidence for the involvement of prostaglandin H synthase in hydroperoxide-dependent oxidation by ram seminal vesicle microsomes. J. Biol. Chem. 258, 6517–6523.

Porter, D.G., Lye, S.J., Bradshaw, J.M.C., & Kendall, J.Z.L. (1981). Relaxin inhibits myometrial activity in the ovariectomized non-pregnant ewe. J. Reprod. Fertil. 61, 409–414.

Rice, G.E. (1995). Glycerophospholipid metabolism and human labor. Reproduction, Fertility and Development 7, 613–622.

Rice, G.E., Wong, M.H., & Thorburn, G.D. (1988). Gestational changes in prostaglandin synthase activity of ovine cotyledonary microsomes. J. Endocrinol. 118, 265–270.

Rollins, T.E., & Smith, W.L. (1980). Subcellular localization of prostaglandin-forming cyclooxygenase in swiss mouse 3T3 fibroblasts by electron microscopic immunocytochemistry. J. Biol. Chem. 255, 4872–4875.

Saphier, P.W., Glynn, B.P., Woods, R.J., Shepherd, D.A.L., Jeacock, M.K., & Lowry, P.J. (1993). Elevated levels of N-terminal pro-opiomelanocortin peptides in fetal sheep plasma may contribute to fetal adrenal gland development and the pre-parturient cortisol surge. Endocrinology 133, 1459–1461.

Stys, S.J., Clewell, W.H., & Meschia, G. (1978). Changes in cervical compliance at parturition independent of uterine activity. Am. J. Obstet. Gynecol. 130, 414–418.

Thorburn, G.D., & Rice, G.E. (1987). Fetal maturation and the timing of parturition: A comparative analysis. Proc. Aust. Physiol. Pharmacol. Soc. 18, 51–68.

Thorburn, G.D., & Liggins, G.C. (1994). Role of the fetal pituitary-adrenal axis and placenta in the initiation of parturition. In: *Marshall's Physiology of Reproduction*, Vol. 3 (Lamming, G.E., Ed.) pp. 1003–1036. Chapman and Hall, London.

Wimsatt, J., Nathanielsz, P.W., & Sirois, J. (1993). Induction of prostaglandin endoperoxide synthase isoform-2 in ovine cotyledonary tissues during late gestation. Endocrinology 133, 1068–1073.

Wood, W.G., Pearce, K., Clegg, J.B., Weatherall, D.J., Robinson, J.S., & Thorburn, G.D. (1976). Switch from fetal to adult haemoglobin synthesis in normal and hypophysectomised sheep. Nature 264, 799–801.

REGULATION OF OVIPOSITION

Frank Hertelendy, Kiyoshi Shimada, Miklós Tóth,
Miklós Molnár, and Kousaku Tanaka

Advances in Organ Biology
Volume 1, pages 31–59.

ISBN: 1-55938-639-8

ABSTRACT

The laying of hard-shelled eggs (oviposition) by all birds and certain reptiles is a pivotal step in reproduction and in the successful survival of these species. The regulation of oviposition, not unlike that of parturition in mammals, is under the influence of multiple factors. Of these, the neuropeptide, arginine vasotocin (AVT), and the primary prostaglandins ($PGF_{2\alpha}$ and PGE_2) play leading roles. In addition, a putative ovarian oviposition-inducing factor may also participate in the timing of oviposition. From extensive *in vivo* and *in vitro* studies a good deal of evidence has accumulated that AVT and prostaglandins promote the contractile activity of the shell gland (uterus) muscle, a function that is coordinated with the opening of the utero-vaginal sphincter under the influence of PGE_2 and the relaxation of the vagina, allowing the egg to be expelled. Results of experiments are presented which indicate that $PGF_{2\alpha}$ and AVT diverge in their cellular and molecular actions, in terms of regulating intracellular calcium ion concentrations that are generally believed to be an essential feature of the contraction/relaxation cycle of uterine smooth muscle. At physiological concentrations, $PGF_{2\alpha}$ promotes Ca^{2+} entry into these cells, whereas AVT activates the phosphoinositide cycle, generating the calcium mobilizer, inositol trisphosphate, as well as enhancing calcium uptake from the extracellular compartment.

I. INTRODUCTION

Although the focus of this volume is on the pregnancy and parturition of eutherian (placental) mammals, oviposition (egg-laying) is an essential feature of reproduction in oviparous reptiles and all birds, representing about three quarters of the species belonging to terrestrial vertebrates. In addition, a subclass of mammals, the monotremes, represented by the duckbilled platypus and the spiny anteater, also reproduce by laying fertilized eggs. The newly hatched offspring subsequently nurse from mammary gland-like structures. Indeed, until the development of the amniotic egg some 250–300 million years ago, viewed by biologists as a landmark in the history of the evolution of terrestrial vertebrates, all embryonic development took place in an external watery environment. The amniotic egg, while offering a similar environment for embryogenesis, an essential attribute throughout the animal kingdom, also provided the nutrients, as well as a protective casing in the form of porous membranes and calcareous shell. This allowed for progressively longer absences from rivers and lakes, resulting in the colonization by reptiles of much of the landmass during the Triassic period (about 200 million years ago), thus replacing the amphibians as the dominant animals on land. The subsequent evolution of birds from reptiles, equipped with a four chambered heart and feathers for flying, added a new dimension to the dispersion of egglaying vertebrates throughout the globe.

In this chapter we shall describe how an egg is made, from ovulation until oviposition, and the various factors believed to be involved in the regulation of these processes. Because much of what we know has been obtained by studying the reproductive biology of the domestic chicken, we have chosen this remarkable animal as the prototypical egglaying model. When appropriate, brief references will be made to other species of birds or certain reptiles.

The reproductive physiology of the domestic fowl has been extensively studied and the interested reader may wish to consult one or more of the following texts and reviews: Bell and Freeman (1971), Nalbandov (1976), Freeman (1983), Cunningham et al. (1984), Cunningham (1987), Etches (1990), and Sharp (1993). Here we only offer a brief description of some of the salient points leading to oviposition.

II. THE REPRODUCTIVE ORGANS AND THEIR FUNCTIONS

The reproductive organs of our model, the domestic hen, consist of a single ovary and oviduct. During the 10–12 days preceding the onset of reproductive activity (i.e., the laying of the first egg), these organs increase in size more than 100-fold and occupy much of the abdominal cavity (Figure 1). Although only the left ovary and oviduct develop during embryogenesis, occasionally a vestigial ovary and oviduct may also be found on the right side.

A. The Ovary

In contrast to mammals, the ovary of the regularly laying hen contains a cluster of follicles of different sizes, reflecting different stages of maturation. The largest one of these (F_1), containing the ovum ("yolk"), will be ovulated first, by mechanisms still largely unknown. Its place will be taken by the second largest follicle (F_2), which will ovulate the next day and so on. This hierarchical arrangement permits the shedding and entry of a single ovum into the oviduct and the subsequent completion of egg formation and oviposition on the following day (see following).

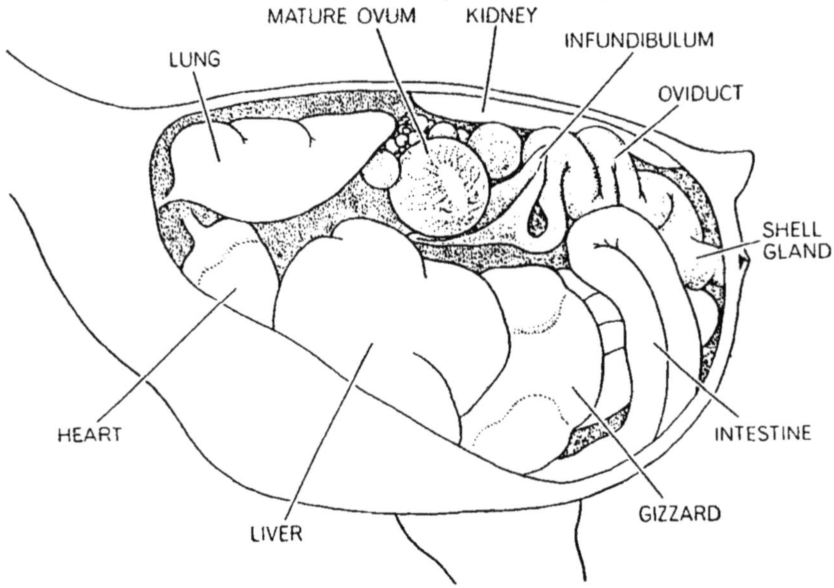

Figure 1. Outline of the reproductive tract of the chicken in relation to the other organs in the body cavity. The single ovary and oviduct are on the hen's left side. (Reprinted with permission from Taylor, T.G., *Scientific American,* March 1970, pp. 88–95.)

Occasionally two follicles may ovulate, giving rise to a double-yolk egg. Ovulation occurs during daylight hours and since it takes a little more than 24 h from ovulation to oviposition, and because rupture of the next follicle takes place about 15–45 min after oviposition, a complete egg cycle requires about 25–26 h. Hence ovulation occurs a little later each day, until the terminal egg in a sequence is ovulated, to be followed by an anovulatory or "pause" day or two. Modern commercial layers, bred for high rates of egg production may have long uninterrupted sequences, lasting for several weeks. Folliculogenesis, as well as ovulation, are controlled by pituitary gonadotropins. Hypophysectomy is rapidly followed by ovarian atresia. Ovulation is preceded by a rise in plasma LH levels, peaking between 4–7 h before ovulation; an injection of LH prior to the endogenous rise will induce premature ovulation. In birds, unlike in mammals, progesterone plays the role of positive signal for LH release, for it can induce LH surge and ovulation when injected at certain times during the normal ovulatory cycle. This role of progesterone is supported by the findings that it is taken up maximally in the anterior pituitary and median eminence 8 h before ovulation. These observations are consistent with the early studies of Ralph and Fraps (1960) who induced premature ovulation in hens by injecting progesterone directly into the hypothalamus or the caudal neostriatum. The dissection of the temporal relationship between the rise in circulating progesterone and LH has not been an easy task, due to the compression of these cyclic hormonal changes into a relatively short period during the daily ovulatory cycle.

B. The Oviduct

The length of the laying hen's oviduct is 55–60 cm long and weighs 45–50 g. It can be divided into morphologically and functionally discrete regions: infundibulum, magnum, isthmus, shell gland (uterus), and vagina (Figure 2). Some investigators identify the region between the shell gland proper and the isthmus as the "tubular shell gland," thus dividing the oviduct into six regions.

Shortly after oviposition has taken place, unless it was the terminal egg in a sequence, a fresh ovum is released from the largest follicle into the abdominal cavity, where it is engulfed by the funnel-shaped segment of the oviduct, the infundibulum (the functional equivalent of the fimbria of the Fallopian tube in mammals). It is here that the egg is fertilized in hens that have been mated. The yolk passes through this approximately 9 cm long portion of the oviduct in 15–20 min. Following this, the yolk enters the magnum, the longest (approx. 30 cm) and most glandular segment of the oviduct, where the albumen, or egg white, is secreted in about 3 h. During the next one and one-half hours the egg traverses the isthmus portion of the oviduct, where the soft shell membranes are formed. Next, the egg, encased in shell membranes, enters the thick muscular pouch, the shell gland, or uterus where it spends about 20–21 h. During this time fluid is taken up into the egg (the "plumping" process) and the hard, calcareous shell is formed. The average egg shell weighs 5 g, of which 2 g is calcium. Most of the shell deposition is accomplished in about 16 h. Thus, calcium is extracted from the circulation at an average rate of 125 mg/h. Since the total amount of calcium in the circulation is about 25 mg at any one time, the hen removes this amount from the blood every 12 min during the period of active shell formation. Part of this is derived directly from dietary sources, but a significant portion is obtained from the skeleton. Under the regulatory influence of sex hormones, the marrow cavities of most bones of reproducing female

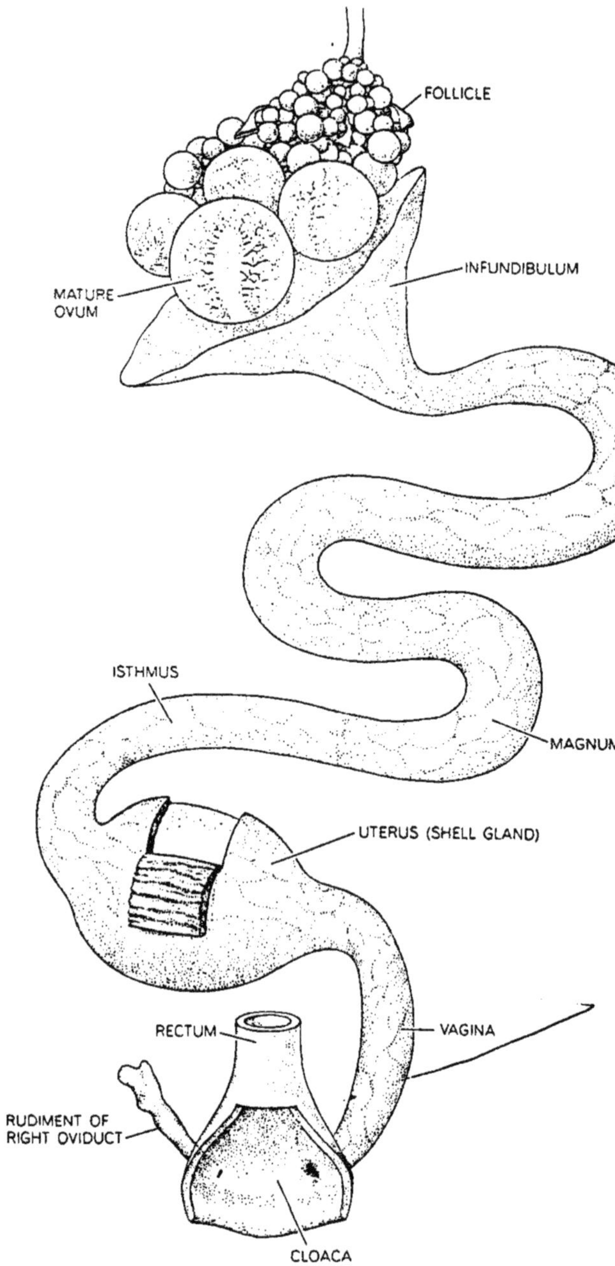

Figure 2. Gross structure of the ovary and oviduct of the chicken. (Reprinted with permission from Taylor, T.G., *Scientific American,* March 1970, pp. 88–95.)

birds are filled with a spongy secondary bone, the so-called "medullary" bone, that has been shown to provide most of the skeletal calcium for shell formation. Not surprisingly, the medullary bone undergoes periods of profound osteoclastic activity during shell calcification, followed by the dominance of osteoblasts, the bone-forming cells. When shell formation has been completed, the egg is pushed into the vagina by the contractile force of the musculature of the uterus and is expelled via the cloaca, in the act collectively referred to as oviposition.

III. REGULATION OF OVIPOSITION

Several putative factors have been implicated in the control of oviposition. These include neurohypophyseal hormones, primarily arginine vasotocin (AVT), sympathomimetic amines, ovarian oviposition inducing factors (OOIF), steroid hormones, and prostaglandins (PG). Evidence will be reviewed to show that, of these, PG and AVT play major roles.

A. Neurohypophyseal Hormones

The hypothalamus of avian, as well as amphibian and reptilian species, elaborates AVT and mesotocin. The structural homology of these neuropeptides with the predominant mammalian neurohypophyseal hormones, oxytocin and vasopressin, is illustrated in Table 1. There is evidence that the avian hormones are also synthesized in the hypothalamus as larger precursor peptides and transported in conjunction with neurophysins along axons to the neurohypophysis. Immunohistochemical studies have identified both vasotocinergic and mesotocinergic neurons in the preoptic region, including the supraoptic nucleus and the paraventricular nucleus.

Table 1. Comparison of Amino Acid Sequences of Mammalian and Avian Oxytocic Peptides

	1	*2*	*3*	*4*	*5*	*6*	*7*	*8*	*9*
Oxytocin	Cys	Tyr	Ile	Gln	Asn	Cys	Pro	Leu	Gly (NH_2)
Arginine vasopressin	Cys	Tyr	Phe	Gln	Asn	Cys	Pro	Arg	Gly (NH_2)
Arginine vasotocin	Cys	Tyr	Ile	Gln	Asn	Cys	Pro	Arg	Gly (NH_2)
Mesotocin	Cys	Tyr	Ile	Gln	Asn	Cys	Pro	Ile	Gly (NH_2)

What is the evidence that AVT is involved in the control of oviposition? The classic study by Riddle (1921) demonstrated that injection of posterior pituitary extract induces premature oviposition in chicken. The subsequent identification of AVT as the principal avian oxytocic hormone (Munsick et al., 1960), and its commercial availability, opened up new avenues of investigation. Thus, it was observed that AVT is more potent than vasopressin or oxytocin in stimulating uterine contractility *in vitro*, as well as inducing oviposition in chickens (Rzasa and Ewy, 1970). Moreover, the efficacy of AVT to induce premature oviposition increases as the time of spontaneous oviposition approaches. An analogous phenomenon, with respect to increased oxytocin sensitivity with approaching parturition in mammals, has been amply documented. However, Olson et al. (1978) did not observe significant differences in contractile responses to AVT when shell gland strips, removed 6 h before oviposition, at oviposition and 6 h after oviposition, were stimulated *in vitro*. A decrease in bioassayable oxytocic substance in the posterior pituitary around the time of oviposition was found to correlate with an increase in circulating levels in the laying hen. Development of sensitive radioimmunoassays allowed the precise measurement of circulating levels of AVT during the laying cycle. Corroborating earlier bioassay data, maximal immunoreactive AVT values were found in plasma at the time of oviposition, declining rapidly to basal levels within 15–30 min (Tanaka et al., 1984). Because of the temporal overlap in the increase in plasma AVT and oviposition, it has not been possible to ascertain whether AVT is responsible for the initiation of oviposition or, as in the case of oxytocin's role in parturition, its release is coupled to a neural reflex activated by the mechanical stretch exerted by the contracting shell gland on the one hand, or, on the other hand, pressure exerted on the vaginal sphincter (the functional homologue of the cervix) by the hard-shelled egg. Yet there is experimental evidence to suggest that plasma AVT levels rise at the time of expected oviposition, even when premature oviposition is induced with PG some hours earlier (Shimada and Saito, 1989). This would suggest that mechanical stretch is not essential for the release of AVT into the circulation and indicates that a "built-in clock" may participate in triggering hormone release from the posterior pituitary.

The source of AVT has also not been clearly established. Even though the available evidence indicates that the release of AVT from the neurohypophysis correlates with elevated plasma levels at the time of oviposition, AVT may also be derived from the ovary, acting in a paracrine fashion on the oviduct. Indeed, both AVT and mesotocin have been identified in extracts of chicken ovaries and ruptured postovulatory follicles. Moreover, Shimada et al. (1987) have found significant levels of immunoreactive AVT in the venous effluent of the largest and second largest preovulatory follicles immediately after oviposition. However, because these values were lower than those measured in peripheral blood collected at the same time, it appears unlikely that AVT from ovarian sources is a major contributor to circulating levels of this neurohormone at the time of oviposition.

B. Ovarian Oviposition-Inducing Factor (OOIF)

The notion that the ovaries are involved in the control of oviposition stems from the observation made by Rothchild and Fraps, more than a half century ago (1944), that the removal of the freshly ruptured follicle (POF_1) caused a striking delay in oviposition from 9 h to 7 days in 31 of 37 such treated hens, whereas sham-operated hens laid at expected times. Subsequently, Tanaka and associates at Kyushu University demonstrated that surgical removal of either the largest preovulatory follicle, or the POF_1 corresponding to the terminal ovulation of a sequence, delayed oviposition until the first ovulation of a new sequence following an anovulatory day. Shimada and colleagues at Nagoya University, using an electrotelemetric device to record uterine activity in the hen, have also demonstrated that the removal or ligation of the largest preovulatory follicle prevented the characteristic pre-ovipository increase in uterine activity. However, if the ligation was performed 30 min before ovulation of the first egg of a sequence, uterine activity was not inhibited, suggesting that if a contractile ovarian factor is released, it must occur 0.5–1 h before the expected ovulation. It may not be a coincidence that prostaglandin levels also peak at this time. The cellular origin of this ovarian factor may be the granulosa, since the scraping away of the granulosa layer, without ligation or removal of the follicle itself, is sufficient to retard oviposition. Moreover, a saline-soluble extract of the ruptured, postovulatory follicle or the preovulatory follicle, removed not earlier than 20 h prior to expected ovulation when injected into normal laying chickens, can induce oviposition prematurely. It has been suggested by the Japanese investigators that this oviposition-inducing follicular factor is a peptide similar in its chemical nature to AVT, vasopressin, or oxytocin. Subsequent efforts by Tanaka's group to isolate the ovarian factor from about 1 Kg of fresh follicular tissue yielded about 4 mg of highly purified active substance, as measured by a hen oxytocic bioassay *in vitro*. The purified product was found to be acidic (isoelectric point about pH 3.5), heat resistant (no loss of bioactivity after heat treatment at 100°C for 10 min), and devoid of prostaglandins. Yet when injected into two laying chickens with soft-shelled eggs in the uterus, it caused immediate oviposition. However, the chemical structure of this putative ovarian oviposition-inducing factor remains unknown.

C. Sympathomimetic Amines

A considerable amount of work has been done on the nervous control of oviposition. Thus, stimulation of the preoptic hypothalamus causes premature voiding of the egg, whereas stimulation of the telencephalon prolongs the time the egg spends in the uterus. Adrenergic mechanisms may also be involved in the activity of the hen's oviduct, since the presence of both alpha and beta receptors has been demonstrated. The stimulation of the latter relaxes the hen's uterus as it does the myometrium, an effect that can be prevented by beta adrenergic blockers. Beta adrenergic receptor-mediated relaxation of uterine smooth muscle has been generally attributed to the activation of the adenylyl cyclase and the intracellular accumulation of cyclic AMP. It was surprising, therefore, to find that intrauterine administration of dibutyryl cyclic AMP to laying hens induced premature oviposition, a phenomenon that might have resulted from the relaxation of the utero-vaginal sphincter and the vaginal segment of the oviduct itself. That the adenylyl cyclase-cyclic AMP signaling pathway is operative in the avian oviduct, including the shell gland, but particularly in the above-mentioned regions of the oviduct was demonstrated by the finding that PGE_1, PGE_2, as well as forskolin alone and, especially, in combination, caused a dose-related increase in cyclic AMP accumulation *in vitro* (Molnár et al., 1987).

In vitro studies on shell gland strips using the mixed α and β agonist, epinephrine, yielded variable responses, whereas the pure β agonist, isoproterenol, consistently relaxed the strips. When intraluminal pressure in the various segments of the laying hen's oviduct was measured, it was observed that norepinephrine promoted contractile activity in all regions, except the shell gland, which relaxed. On the other hand, isoproterenol caused only relaxation throughout the oviduct, an effect that was blocked by propranolol. Despite these, and some other similar observations, it seems unlikely that sympathetic mechanisms play a decisive role in the regulation of oviposition, since transection of the sympathetic nerves leading to the shell gland had no effect on egg laying.

It appears more likely, therefore, that endocrine and paracrine factors may play an important, functional role in the regulation of oviposition.

D. Steroid Hormones

Whereas steroid hormones, primarily estrogen and progesterone, have a profound regulatory influence on parturition, their role in avian oviposition is obscure, even though the chicken oviduct has served as a widely used model to study the mode of action of estrogen and progesterone. It has been suggested that the progesterone-dominated oviduct of the hen, unlike the mammalian uterus, is active. Other studies, however, show a significant fall in plasma, as well as follicular progesterone levels shortly before oviposition. Clearly, new efforts are necessary to examine, both on organ, cellular, and molecular levels, the potential role of female sex steroids in the regulation of oviposition.

IV. ROLE OF PROSTAGLANDINS IN OVIPOSITION

Prostaglandins belong to a group of compounds, the eicosanoids, a populous family of oxygenated and conjugated derivatives of the 20 carbon polyunsaturated essential fatty acid, arachidonic acid (AA). Prostaglandins are synthesized via the cyclooxygenase pathway by the enzyme, prostaglandin endoperoxide H_2 synthase, (cyclooxygenase), of which two isoforms have been identified, the constitutively expressed type 1 (COX-1) and the inducible type 2 (COX-2). In addition, AA can be metabolized by the lipoxygenase pathway to various leukotrienes and other products, or via the epoxygenase route by cytochrome P_{450} enzymes to various epoxy fatty acids (Figure 3). The view has been generally held that the release of AA from phospholipids represents the rate-limiting step in the biosynthesis of prostaglandins. Although AA can be released by the activation of various phospholipases (see Figure 4), growing evidence indicates that a cytosolic isoform of phospholipase A_2 ($cPLA_2$) may play a key role in agonist-activated release of AA in a variety of cells, by a mechanism involving the phosphorylation of this enzyme by mitogen-activated protein kinases.

Of all the putative factors implicated in the regulation of oviposition, the physiological role and mode of action of the primary prostaglandins E_2 and $F_{2\alpha}$ have been the most extensively studied. Below we shall review this evidence, which leads us to propose that these prostaglandins are an

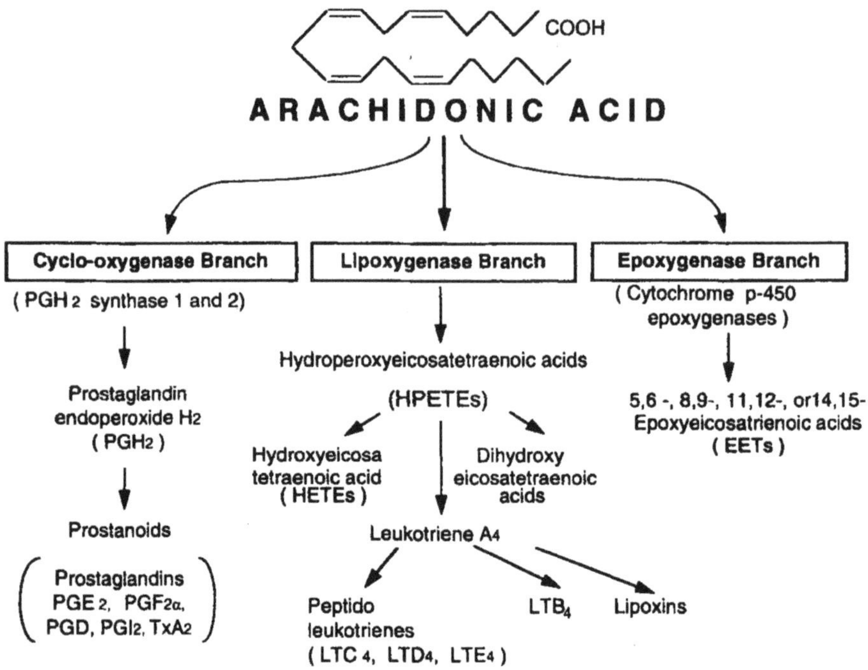

Figure 3. The arachidonic acid (AA) cascade. Note that the primary prostaglandins are generated via the cyclooxygenase branch of AA metabolism. (Reprinted with permission from Hertelendy, F., and Tóth, M. (1993). Proc. Zool. Soc. Calcutta, Haldane Comm. 195–216.)

essential component of the physiology of avian and, possibly, reptilian (Guillette, 1990) oviposition.

A. *In Vivo* Effects

Early experiments have demonstrated that the intrauterine administration of small amounts of these prostaglandins induced premature egg-laying within minutes in the Japanese quail, domestic chicken, and the turkey hen (Table 2). (These observations have subsequently been confirmed in a number of laboratories in the United States, as well as in Europe and Japan.) Moreover, intrauterine injection of AA, the principal precursor of prostaglandins and other prostanoids, as well as phospholipase A, the enzyme most responsible for the liberation of free AA from phospholipid stores, also induced oviposition. These effects were inhibited by adminstration of the cyclooxygenase blockers, indomethacin, and

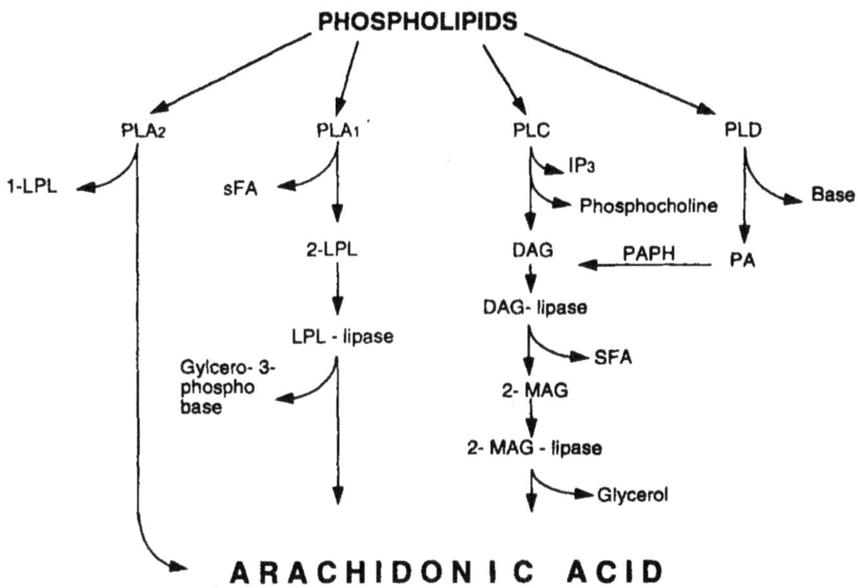

Figure 4. Known pathways of arachidonic acid release from phospholipids. PL: phospholipase; PAPH: phosphatidic acid phosphohydrolase; DAG: diacylglycerol; LPL: lysophospholipid; MAG: monoacylglycerol; SFA: saturated fatty acid. (Reprinted with permission from Hertelendy, F., and Tóth, M. (1993). Proc. Zool. Soc. Calcutta, Haldane Comm. 195–216.)

Table 2. Prostaglandins were Injected Directly into the Uterus Several Hours before Expected Oviposition

	Induction of Oviposition by Prostaglandins				
Species	*Compound*	*Dose (mg)*	*# of Tests*	*Successful Induction*	*Average Induction Time (min)*
Quail	PGE_1	0.01	7	5	
		0.1	19	15	3.0
	PGE_2	0.01	8	2	5.5
		0.1	9	5	2.0
		1.0	4	4	4.0
	$PGF_{2\alpha}$	1.0	9	7	21.5
Chicken	PGE_1	0.05	7	4	9.5
		0.25	39	39	3.9
	PGE_2	0.05	5	2	10.5
		0.1	5	5	5.2
	$PGF_{2\alpha}$	0.1	8	3	4.7
		1.0	6	6	3.2
Turkey	PGE_1	1.0	5	5	7.0
	PGE_2	1.0	5	5	9.0
	$PGF_{2\alpha}$	1.0	4	1	7.0
		5.0	4	4	7.5

Source: Hertelendy, F. In: *Avian Endocrinology* (Epple, A., & Stetson, M.H., Eds), pp. 445–479.

aspirin, indicating the presence and operation of the enzymatic machinery for prostaglandin synthesis in the avian uterus.

Additional experiments provided further evidence in support of a key role for prostaglandins in the physiology of oviposition. First, an oral administration of indomethacin several hours before expected spontaneous oviposition caused a retention of the egg in the uterus for an additional 12 h, or about 50% longer than the normal length of time an egg spends in the shell gland, while control hens receiving placebo laid at the expected time. Such treatment with indomethacin reduced plasma PGE levels by approximately 90%. Furthermore, injection of prostaglandins E or $F_{2\alpha}$ in indomethacin-blocked chickens invariably induced oviposition. It is of interest to note that in these and other *in vivo* studies, the E series of prostaglandins were found to be markedly more potent inducers of oviposition than $PGF_{2\alpha}$, due perhaps to their contractile effect on uterine smooth muscle, combined with their relaxant action on the uterovaginal sphincter and vagina. These properties of PGE_2 are believed to underlie its efficacy in the induction of labor at term or interruption of pregnancy, in combination with RU486, in humans.

Second, neutralization of endogenous prostaglandin with PGE antiserum, raised in a goat, inhibited spontaneous oviposition, by a duration similar to that observed with indomethacin, whereas injection of normal goat serum had no effect. Third, plasma levels of PGFM, the stable metabolite of $PGF_{2\alpha}$, rise significantly at the time of oviposition, both at midsequence and at the terminal oviposition of a laying sequence (Figure 5). Significantly, PGFM levels increased at the time of expected oviposition, even when such was induced prematurely by intravenous injection of AVT several hours earlier. Thus, the rise in prostaglandin levels cannot be attributed to physical stretch or increased muscular activity, factors that have been implicated in conjunction with mammalian parturition. However, the potential influence of the presence of a hard egg in the shell gland cannot be completely ruled out, because the efficacy of exogenous prostaglandins to induce oviposition increases during the laying cycle as the time of expected oviposition approaches. Similar findings have also been reported for the efficacy of AVT. Interestingly, the sensitivity of the mammalian uterus to uterotonic agents also increases progressively with approaching parturition. In sub-primate species exhibiting progesterone withdrawal before the onset of parturition, this phenomenon has been explained by the "progesterone block" theory (Csapo, 1956). Fourth, electrophysiological data, obtained *in vivo* from chronically instrumented chickens, demonstrating increased electrical activity at the time

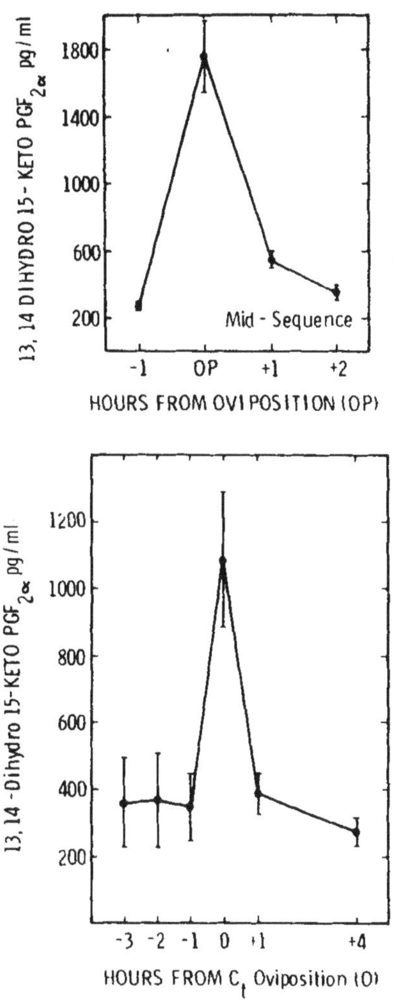

Figure 5. Plasma PGFM levels around the time of midsequence oviposition (upper panel) and during terminal oviposition (Ct), when oviposition is not followed by ovulation (lower panel). The PGFM values at oviposition are significantly greater compared to the other points on the graphs. (Reprinted with permission from Olson, D.M., and Hertelendy, F. (1981). Biol. Reprod. 24, 496–504.)

of oviposition, regardless of whether such occurred spontaneously or following premature induction with AVT, provided good correlation with concomitant changes with prostaglandin levels (Shimada et al., 1984).

In another study, prompted by the possibility that prostaglandins, when administered locally, may act as detergents on the surrounding musculature, we investigated the effects of a nonionic detergent, Triton X-100, as well as deoxycholate, on oviposition in chickens (Hertelendy et al., 1979). Intrauterine injection of 0.5 ml of either a 0.2% solution of Triton X-100 or a 0.2% solution of deoxycholate, 3–5 h before the expected time of spontaneous oviposition, induced premature egg laying in 10/10 hens within a mean time of 7.8 min, or 5/5 hens within 13.2 min, respectively. None of the control birds which received the same volume of saline laid before the predicted time of oviposition, based on carefully kept records. However, when hens were pretreated with indomethacin several hours prior to the injection of detergents, none of the 10 birds tested laid. In fact, eggs were retained by an average time of 7.6 h beyond the expected time of spontaneous oviposition. Similar results were obtained when hens were pretreated with cortisol, another known inhibitor of prostaglandin synthesis. Once again, the oviposition inhibitory effect of indomethacin in detergent-treated hens could be completely reversed by administration of exogenous PGE_2. These experiments indicated that the action of detergents was mediated by PG release, adding further credence to the notion that locally produced prostaglandins have important functional roles in initiating and completing oviposition. Although the mechanism by which detergents promote prostaglandin output *in vivo* is unknown, *in vitro* evidence indicates that Triton X-100 affects phospholipid metabolism in human decidua and primordial placenta preparations (Tóth et al., 1987).

Taken together, these *in vivo* studies, using various experimental approaches, have established the premier role for prostaglandins in the process of oviposition.

B. Sources of Prostaglandins

Ovarian Sources

There is considerable evidence that the rise in circulating prostaglandin levels around the time of oviposition is derived from ovarian follicles. Day and Nalbandov (1977) reported a fourfold increase in PGF concentration in the largest preovulatory follicle and a 100-fold increase in the ruptured postovulatory follicle 1–2 h before oviposition. This rise in PGF levels, as well as oviposition, were inhibited by prior treatment with indomethacin. The authors suggested that PGF, originating from the largest postovulatory follicle, may play a physiological role in controlling uterine contractions and oviposition. Subsequent experiments by other investigators (Hammond et al., 1980), however, observed a 20-fold increase in PGF concentration in the largest preovulatory follicle at the time of oviposition, but not in the two largest postovulatory follicles. Similarly, Shimada et al. (1984), who measured the levels of PGF and several other prostaglandins in the peripheral plasma, as well as in follicular venous effluent and uterine tissue, found a 150-fold increase in PGF and PGFM in plasma samples collected directly from the largest preovulatory follicle, corresponding to maximal uterine contractions preceding ovulation/oviposition. Interestingly, steroid hormone levels (estrogen and progesterone) were also much higher in F_1 plasma, reflecting, probably, the preovulatory rise in gonadotropin levels. These observations were extended by Canadian investigators (Etches et al., 1990) who reported that F_1 follicles *in vitro* and granulosa cells isolated from such follicles released significantly more $PGF_{2\alpha}$ into the incubation medium than smaller, less mature follicles. Interestingly, when granulosa cells collected 6 h before expected spontaneous oviposition were stimulated, first with LH and then with A23187, $PGF_{2\alpha}$ release was increased by 15–20-fold. It would appear, therefore, that LH may regulate not only ovarian steroidogenesis, but may also activate COX-1 and/or induce COX-2 in granulosa cells. How ovarian prostaglandins are transferred to the oviduct remains uncertain.

Uterine Sources

The original observations by Hertelendy (1972), demonstrating that arachidonic acid can induce oviposition that can be blocked by inhibitors of prostaglandin synthesis, indicated that uterine tissues can produce prostaglandins locally. This notion was confirmed in a series of biochemical studies using uterine tissue preparations (Asbóth et al., 1983, 1985; Tóth et al., 1983). Asbóth et al. (1983) evaluated prostaglandin production in a microsomal fraction (90,000 x g pellet of a 600 x g supernatant) prepared separately from the muscular ("myometrium") and glandular ("endometrium") portion of the hen uterus. Unlike tissue fragments, which were found to metabolize AA in the absence of added cofactor, none of the microsomal preparations were able to produce prostanoids. However, when epinephrine or tryptophan were included in the reaction mixture, the conversion of AA to TxB_2, PGE_2 and $PGF_{2\alpha}$ was markedly enhanced. Of particular interest was the presence of TxB_2 as the major metabolite of AA, that could not be attributed to contamination with thrombocytes. Moreover, addition of reduced glutathione

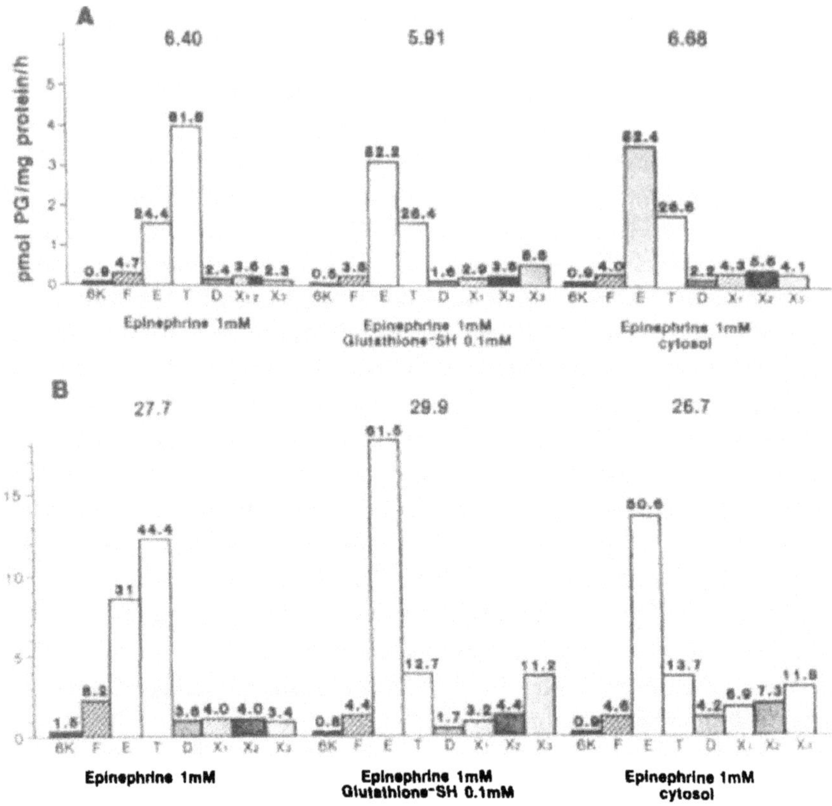

Figure 6. Effect of cofactors and cytosol on prostanoid synthesis from radioactive arachidonic acid in uterine (A) and vaginal (B) membrane preparations. Numbers on top of bars give percentage distribution of radioactivity. Numbers in bold type in upper part of figure show total prostanoid synthesis in pmol/mg/protein/h. 6K, 6-Keto-$PGF_{1\alpha}$; F, $PGF_{2\alpha}$; E, PGE_2; T, TxB_2; D, PGD_2; $X_{1,2,3}$, unidentified products. (Reprinted with permission from Asbóth, G., et al. (1985). Am. J. Physiol. 248, E80–E88.)

(GSH) inhibited TxB_2 formation, while enhancing the synthesis of PGE_2. This inhibition by GSH and some other reducing agents did not appear to be related to a decrease in oxygen concentration in the medium, since reducing agents that do not auto-oxidize readily (mercaptoethanol, dithiothreitol) were found by these investigators to be even stronger inhibitors than sodium dithionate, which can react with oxygen directly. It was postulated, therefore, that the site of inhibition was at the cyclo-oxygenase step.

When the prostaglandin synthesizing capacity of a crude membrane preparation derived from the myometrium and endometrium was compared, the smooth muscle preparation was found to have a 2.5-fold greater capacity to synthesize PGE_2, suggesting that the myometrium possesses a special ability to amplify its own contractile activity by increased production of prostaglandins that may act in a paracrine or intracrine fashion (Tóth et al., 1983). In a subsequent study, the same investigators (Asbóth et al., 1985) reported that the vaginal portion of the oviduct was metabolically highly active, producing four times more prostanoids from [^{3}H]AA than a similar preparation from the shell gland. Once again GSH shifted the metabolic pathway from TxB_2 to PGE_2 (Figure 6).

In summary, these studies have established that the avian uterus/shell gland is an active, prostaglandin-synthesizing organ, capable of an "autocatalytic" function by producing the bioactive prostanoids locally, which may amplify the signal, culminating in coordinated muscular contraction and oviposition.

V. MECHANISM OF ACTIONS OF PROSTAGLANDINS AND ARGININE VASOTOCIN

A. Interactions *In Vivo* and *In Vitro*

The inhibition of oxytocin-induced oviposition in the Coturnix quail by pretreatment with indomethacin or eicosatetraynoic acid, two structurally diverse blockers of prostaglandin biosynthesis, suggested that the uterotonic peptide acts, at least in part, by promoting prostaglandin synthesis (Hertelendy, 1973). However, using longitudinal strips of shell gland muscle of laying hens, we observed that, whereas indomethacin suppressed AA-induced contractility, it did not reduce significantly AVT-dependent tension. It did, however, delay the development of AVT-induced half-maximal, as well as maximal, tension. It would appear, therefore, that prostaglandins are not mediating AVT-and oxytocin-promoted uterine activity. Rather, the two agonists, acting in tandem, intensify the contractile responses of the uterine smooth muscle.

Olson and Hertelendy (1983) examined, in some detail, shell gland muscle contractility *in vitro*, by comparing the effects of $PGF_{2\alpha}$ and AVT. They found that the dose-related tension generated by both of these agonists was significantly suppressed by calcium channel blockers and completely abrogated by omission of calcium in the tissue bath, com-

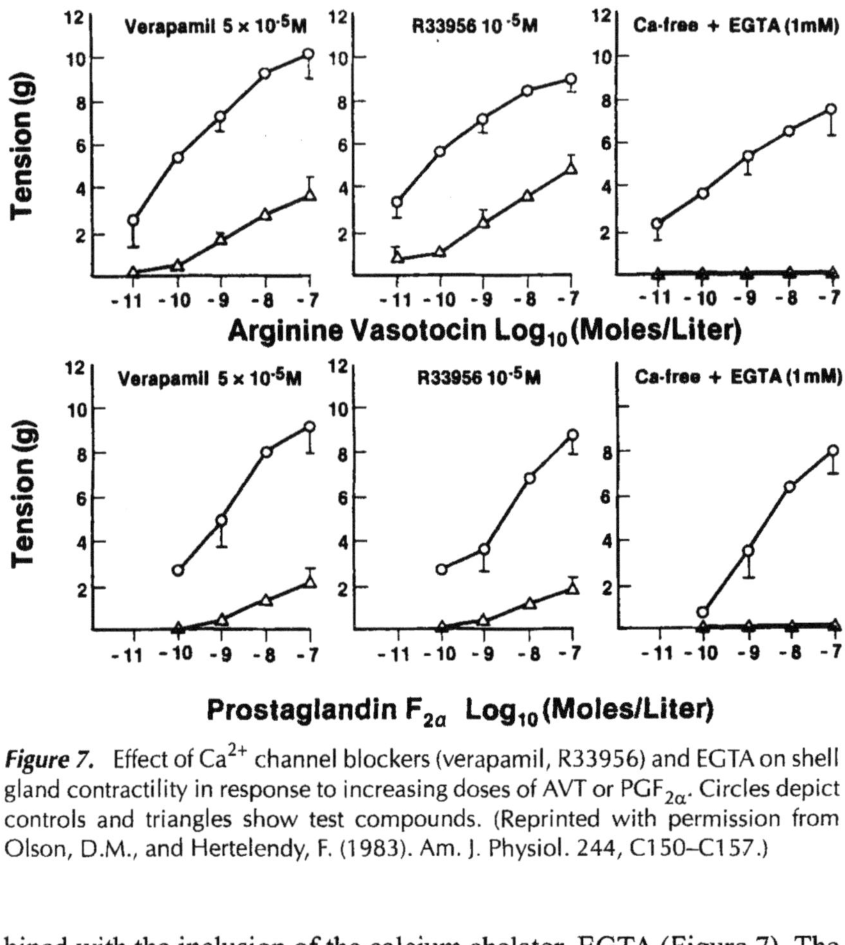

Figure 7. Effect of Ca^{2+} channel blockers (verapamil, R33956) and EGTA on shell gland contractility in response to increasing doses of AVT or $PGF_{2\alpha}$. Circles depict controls and triangles show test compounds. (Reprinted with permission from Olson, D.M., and Hertelendy, F. (1983). Am. J. Physiol. 244, C150–C157.)

bined with the inclusion of the calcium chelator, EGTA (Figure 7). The dependence of tension generation on progressively increasing the concentration of Ca^{2+} in the tissue bath was also clearly demonstrated. On the other hand, marked differences were observed in the way in which varying the extracellular Na^{+} concentrations affected $PGF_{2\alpha}$ and AVT-induced contractile activity. Whereas $PGF_{2\alpha}$ promoted activity was greatly diminished at low Na^{+} concentrations, AVT activity was unaffected (Figure 8). Moreover, depolarization with 20 mM KCl potentiated $PGF_{2\alpha}$ provoked tension, without it having any influence on AVT-induced responses. These and other observations prompted these authors to propose a mechanism by which both $PGF_{2\alpha}$ and AVT interact with discrete sarcolemmal components, triggering an enhanced influx of extracellular Ca^{2+}. $PGF_{2\alpha}$ could affect this process indirectly by altering

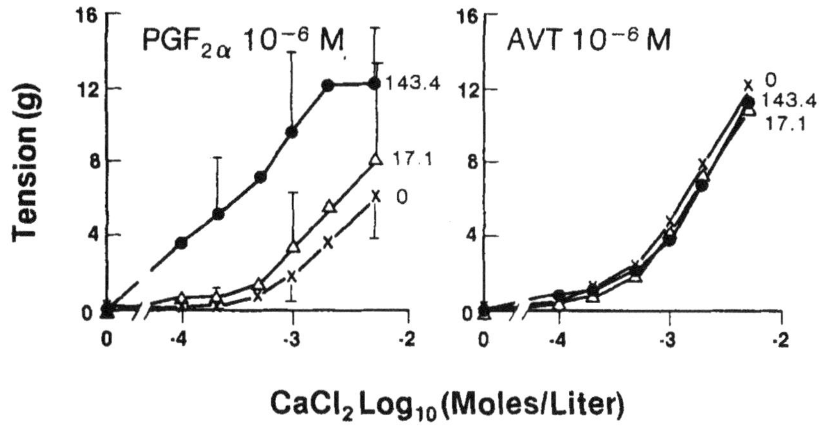

Figure 8. Effect of extracellular Ca^{+} concentration on $PGF_{2\alpha}$ and AVT-induced increases in tension of isolated shell gland strips at increasing concentrations of extracellular Ca^{2+}. Number next to symbols denotes $[Na^{+}]$ in mmol/L. (Reprinted with permission from Olson, D.M., and Hertelendy, F. (1983). Am. J. Physiol. 244, C150–C157.)

Na^{+} permeability, leading to depolarization, which would open voltage-gated Ca^{2+} channels, whereas AVT may directly bring about membrane depolarization, by increasing the permeability of the sarcolemma to Ca^{2+}.

B. Signal Transduction

It is generally accepted that the first step in signal transduction by an extracellular agonist is a specific, high affinity association with its receptor in the plasma membrane. Using arginine vasopressin (AVP) as the isotopically labeled ligand, Koike et al. (1988) were able to demonstrate a single class of limited number of binding sites in a shell gland membrane preparation. AVP and AVT were about equipotent in displacing $[^{3}H]$AVP, whereas mesotocin was ineffective. A more detailed investigation on prostaglandin binding was carried out by Tóth et al. (1979) and Asbóth et al. (1985). These studies revealed a specific binding of $PGF_{2\alpha}$ to a membrane fraction prepared from shell gland muscle of laying hens. The uptake of labeled ligand was concentration, time, and temperature-dependent. Scatchard analysis of $PGF_{2\alpha}$ binding indicated a single class of binding sites with an apparent K_d of 6×10^{-7} M and binding capacity of about 1 pmol/mg protein. Significantly a good correlation was observed between the parameters of $[^{3}H]PGF_{2\alpha}$ binding and contractile responses of uterine strips. Experiments (Tóth et al., 1981) using differential centrifugation to obtain various fractions from shell gland muscle homogenate, demonstrated that $PGF_{2\alpha}$ binding was highest in those fractions which were enriched the most in 5′-nucleotidase, Ca^{2+}-ATPase, and Mg^{2+}- (Na^{+}+K^{+}) ATPase, enzymes generally used as markers for cell membranes. These experiments support the view that specific interaction of prostaglandins with discrete receptors in the sarcolemma is an integral component of the chain of events leading to enhanced uterine contractility.

Specific binding of PGE_2 was also demonstrated in membrane preparations obtained from both shell gland and vaginal segments of the hen oviduct (Figure 9). However, whereas binding characteristics indicated the presence of a single class of high affinity (apparent K_d about 1 nM), low capacity (B_{max} about 15 fmol PGE_2/mg protein) binding sites in the vaginal preparation, the shell gland exhibited two; a high and low affinity component (Figure 10). The authors speculated that these differences in binding properties relate to the different functions of these discrete regions of the oviduct. Whereas PGE_2 and $PGF_{2\alpha}$ are the dominant uterotonic prostaglandins in the uterus, PGE_2 is concerned with promoting relaxation of the vagina.

Since prostaglandin and oxytocin bind to receptors possessing the typical seven membrane-spanning domains, ligand-receptor-generated signals are transduced by GTP binding proteins. In turn, G-proteins regulate the function of various enzymes (e.g., adenylyl cyclase, phospholipases), as well as ion channels modulating the levels of intracellular signaling molecules, such as cyclic AMP, inositol phosphates, Ca^{2+}, and

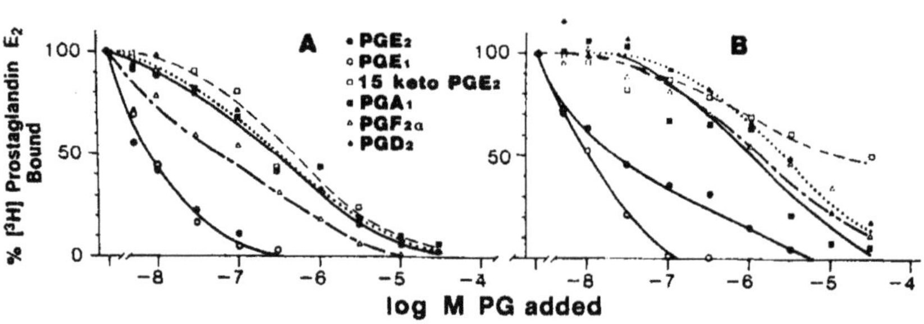

Figure 9. Specificity of PGE_2 binding in uterine (A), and vaginal (B) membrane preparation. Curves depict the displacement of [^{3} H]PGE_2 by increasing concentrations of unlabeled ligands. (Reprinted with permission from Asbóth, G., et al. (1985). Am. J. Physiol. 248, E80– E88.)

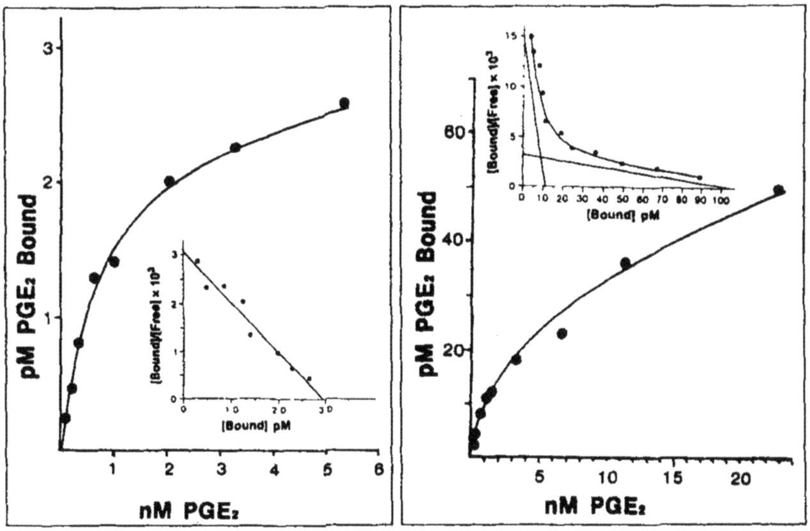

Figure 10. Specific binding of PGE_2 to vaginal (A), and uterine (B) preparations. Insets show Scatchard plots constructed from binding data. (Reprinted with permission from Asbóth, G., et al. (1985). Am. J. Physiol. 248, E80–E88.)

so on. Using primary cultures of shell gland smooth muscle cells, Molnár and Hertelendy (1990) investigated $PGF_{2\alpha}$ and AVT signaling. Because of the central role attributed to intracellular Ca^{2+} concentrations ($[Ca^{2+}]_i$), these studies were directed to examine the mechanisms by which these natural agonists modulate calcium fluxes.

When permeabilized cells were loaded with $^{45}Ca^{2+}$ to isotopic equilibrium in the presence of ruthenium red, to prevent uptake by mitochondria, and were then stimulated with $PGF_{2\alpha}$, AVT, or inositol trisphosphate [Ins(1,4,5)P_3], the agent responsible for Ca^{2+} mobilization by agonists that activate the phosphoinositide cycle, there was a rapid biphasic efflux of ^{45}Ca (Figure 11). The first phase peaked in 60 seconds, whereas the second peak was observed at 5 min after the addition of agonists. AVT proved to be the most potent, provoking half-maximal $^{45}Ca^{2+}$ efflux at 1.2×10^{-9} M versus 9.8×10^{-9} M for $PGF_{2\alpha}$ and 7.5×10^{-6} M for Ins(1,4,5)P_3 during the rapid phase. However, whereas an inhibitor of phospholipase C significantly attenuated AVT-induced $^{45}Ca^{2+}$ mobilization from non-mitochondrial pools, the action of $PGF_{2\alpha}$ (and, of course, that of Ins[1,4,5]P_3) was not affected. Moreover, in the absence of extracellular Ca^{2+}, $PGF_{2\alpha}$ was unable to generate a calcium signal in

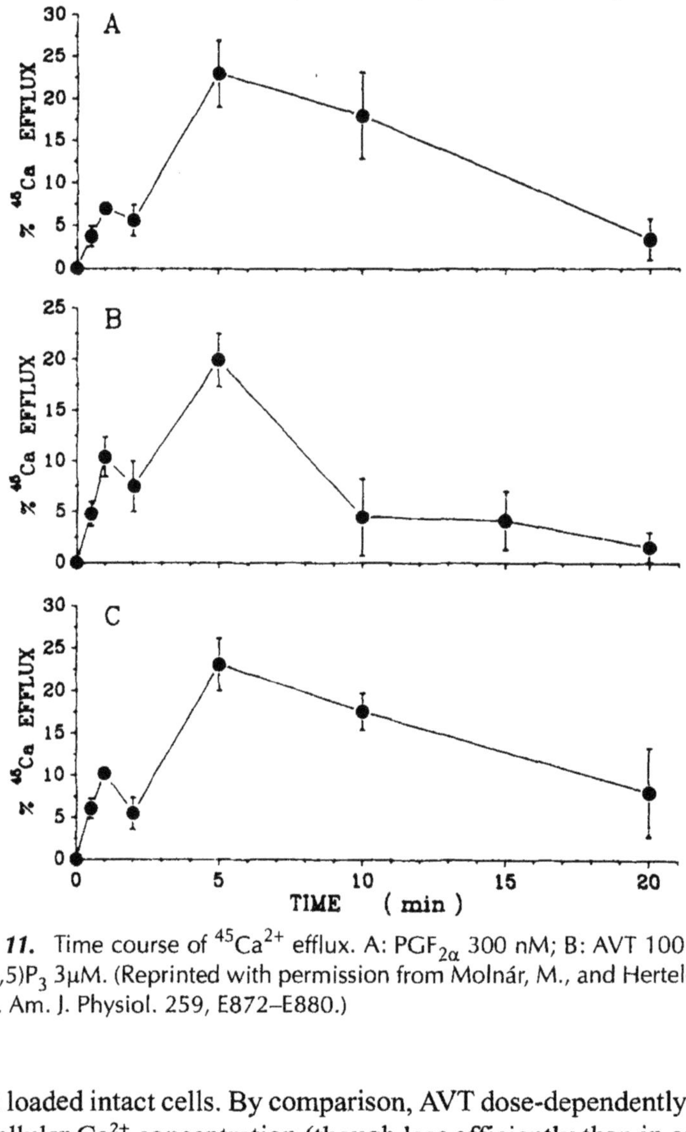

Figure 11. Time course of $^{45}Ca^{2+}$ efflux. A: $PGF_{2\alpha}$ 300 nM; B: AVT 100 nM; C: $Ins(1,4,5)P_3$ 3μM. (Reprinted with permission from Molnár, M., and Hertelendy, F. (1990). Am. J. Physiol. 259, E872–E880.)

fura-2 loaded intact cells. By comparison, AVT dose-dependently raised intracellular Ca^{2+} concentration (though less efficiently than in calcium containing medium) even when cells were suspended in a Ca^{2+}-free medium (Figure 12).

These experiments suggested a different signaling pathway for $PGF_{2\alpha}$ and AVT. The finding that $PGF_{2\alpha}$, at concentrations which maximally raised $[Ca^{2+}]_i$, was ineffective in promoting the release of inositol phosphate from isotopically labeled phosphoinositides in intact cell cultures, while the potency of AVT closely correlated with its efficacy to raise $[Ca^{2+}]_i$, added further support to this notion (Figure 13). Only when

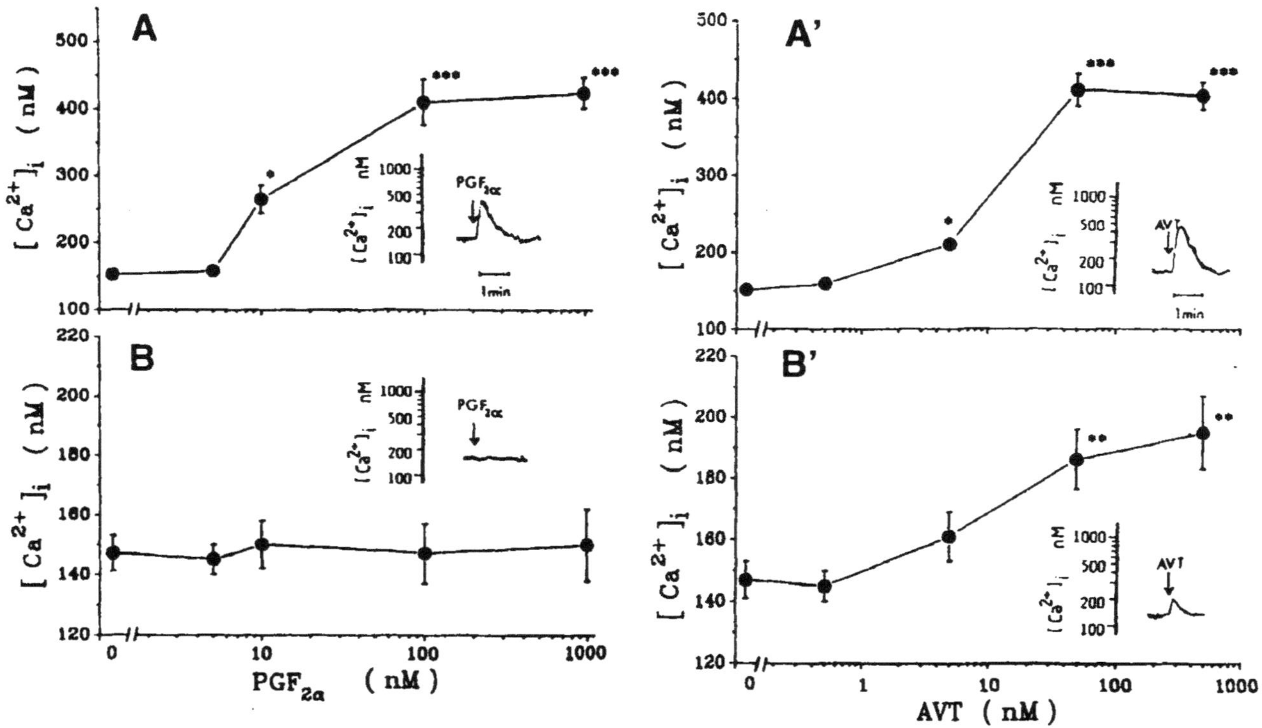

Figure 12. Effect of PGF$_{2\alpha}$ and AVT on intracellular Ca^{2+} concentrations ($[Ca^{2+}]_i$) in isolated smooth muscle cells loaded with fura-2. A: PGF$_{2\alpha}$ + extracellular Ca^{2+}; B: PGF$_{2\alpha}$ without Ca^{2+}; A':AVT + Ca^{2+}; B': AVT without Ca^{2+}. Insets show typical tracings of Ca^{2+} signals in response to 1 μM PGF$_{2\alpha}$ or 100 nM AVT. (Reprinted with permission from Molnár, M., and Hertelendy, F. (1990). Am. J. Physiol. 259, E872–E880.)

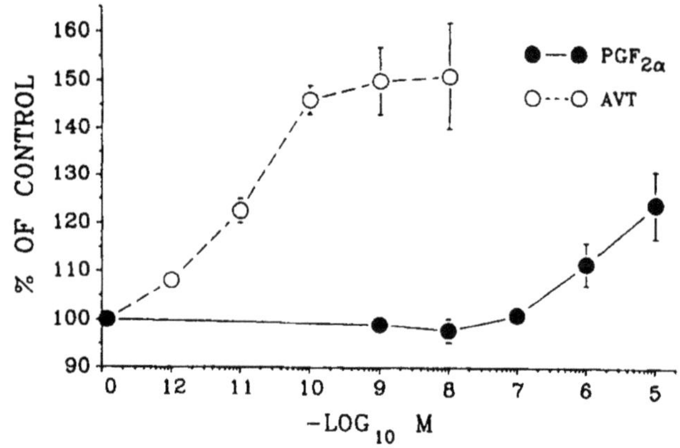

Figure 13. Dose response of AVT and $PGF_{2\alpha}$-stimulated phosphoinositide hydrolysis, as measured by total inositol phosphate production in cultured shell gland smooth muscle cells. (Reprinted with permission from Molnár, M., and Hertelendy, F. (1990). Am. J. Physiol. 259, E872–E880.)

$PGF_{2\alpha}$ concentration was raised to pharmacological levels (1–10 μM) was there a measurable increase in inositol phosphate production. Finally, the calcium channel blocker, verapimil, had no effect on AVT-stimulated inositol phosphate generation, while high concentrations (10 μM) of $PGF_{2\alpha}$-provoked responses were abolished (Figure 14).

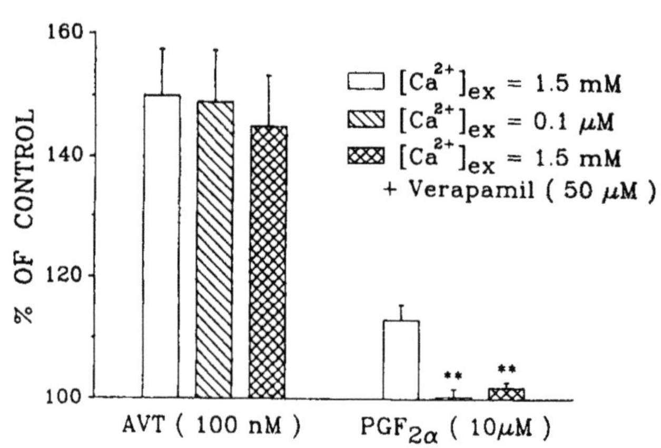

Figure 14. Effect of extracellular Ca^{2+} ($[Ca^{2+}]_e$) and verapamil on AVT and $PGF_{2\alpha}$-stimulated inositol phosphate production in D-myo-[^{3}H]inositol-labeled smooth muscle cells in monolayer cultures. (Reprinted with permission from Molnár, M., and Hertelendy, F. (1990). Am. J. Physiol. 259, E872–E880.)

Taken together, these studies demonstrated that both $PGF_{2\alpha}$ and AVT can influence the handling of Ca^{2+} by smooth muscle cells of the avian shell gland *in vitro.* Moreover, these results also suggest that $PGF_{2\alpha}$ acts primarily by promoting influx of extracellular Ca^{2+}, which may in turn amplify the signal by activating phospholipase A_2 and the release of AA that has been shown to act by itself as a calcium-mobilizing agent (Hertelendy et al., 1992, 1995). AVT, on the other hand, in addition to promoting Ca^{2+} influx through the sarcolemma, activates the phosphoinositide cycle, generating Ins(1,4,5)P_3, which releases Ca^{2+} from the sarcoplasmic reticulum. A complementary signaling pathway for AVT may also be operative in these cells, in view of the recent demonstration that oxytocin activates a mitogen-activated protein kinase (MAPK) in human myometrial cells (Ohmichi et al., 1995), even though the functional role of the MAPK cascade vis-à-vis smooth muscle contraction remains obscure. Nor is there any information available on the intervening steps in the signal transduction pathway between G-protein-coupled receptor activation and MAPK activation in uterine smooth muscle cells. Clearly, this may become one of the exciting new areas of investigation, aimed at understanding the intricacies of the signaling pathways initiated by oxytocic peptides, prostaglandins, and other hormonal substances affecting myometrial function, both in avian and mammalian species.

REFERENCES

Asbóth, G., Tóth, M., & Hertelendy, F. (1983). Conversion of arachidonic acid to prostanoids in the avian uterus. Biochim. Biophys. Acta 750, 481–489.

Asbóth, G., Todd, H., Tóth, M., & Hertelendy, F. (1985). PGE_2 binding, synthesis and distribution in the hen oviduct. Am. J. Physiol. 248, E80–E88.

Bell, D.J., & Freeman, B.M. (1971). *Physiology and Biochemistry of the Domestic Fowl*, Vol. 1–3. Academic Press, London.

Csapo, A.I. (1956). Progesterone "block." Amer. J. Anatomy 88, 273–292.

Cunningham, F.J. (1987). The ovulatory cycle of the hen. *Oxford Reviews of Reproductive Biology* 9, 96–136.

Cunningham, F.J., Lake, P.E., & Hewitt, D. (Eds.). (1984). *Reproductive Biology of Poultry*. Longman, Harlow, U.K.

Day, S.L., & Nalbandov, A.V. (1977). Presence of prostaglandin F (PGF) in hen follicles and its physiological role in ovulation and oviposition. Biol. Reprod. 16, 486–494.

Etches, R.J. (1990). The ovulatory cycle of the hen. Crit. Rev. Poultry Biol. 2, 103–110.

Etches, R.J., Kelly, J.D., Anderson-Langmuir, C.E., & Olson, D.M. (1990). Prostaglandin production by the largest preovulatory follicles in the domestic hen (*Gallus domesticus*). Biol. Reprod. 43, 378–384.

Freeman, B.M. (Ed.). (1983). *Physiology and Biochemistry of the Domestic Fowl*, Vol. 4–5. Academic Press, London.

Guillette, L.J., Jr. (1990). In: *Progress in Comparative Endocrinology* (Epple, A., Scanes, C.G., & Stetson, M.H., Eds.) pp. 603–607. Wiley-Liss, New York.

Hammond, R.W., Olson, D.M., Frenkel, R.B., Biellier, H.V., & Hertelendy, F. (1980). Prostaglandins and steroid hormones in plasma and ovarian follicles during the ovulation cycle of the domestic hen (Gallus domesticus). Gen. Comp. Endocrinol. 42, 195–202.

Hertelendy, F. (1972). Prostaglandin-induced premature oviposition in the coturnix quail. Prostaglandins 2, 269–279.

Hertelendy, F. (1973). Block of oxytocin-induced parturition and oviposition by prostaglandin inhibitors. Life Sci. 13, 1581–1589.

Hertelendy, F., Tóth, M., & Biellier, H.V. (1979). Detergent-induced oviposition is mediated by prostaglandin. Prostaglandins and Medicine 2, 217–223.

Hertelendy, F., Molnár, M., & Jamaluddin, M. (1992). Dual action of arachidonic acid on calcium mobilization in avian granulosa cells. Mol. Cell. Endocrinol. 83, 173–181.

Hertelendy, F., Molnár, M., & Rigó, J., Jr. (1995). Proposed signaling role of arachidonic acid in human myometrium. Mol. Cell. Endocrinol. 110, 113–118.

Koike, T.I., Shimada, K., & Cornett, L.E. (1988). Plasma levels of immunoreactive mesotocin and vasotocin during oviposition in chickens: Relationship to oxytocic action of the peptides *in vitro* and peptide interaction with myometrial binding sites. Gen. Comp. Endocrinol. 70, 119–126.

Molnár, M., & Hertelendy, F. (1990). Ca^{2+} release and $InsP_3$ production in avian uterine cells: Effects of $PGF_{2\alpha}$ and AVT. Am. J. Physiol. 259, E872–E880.

Molnár, M., Asem, E.K., & Hertelendy, F. (1987). Differential effects of $PGF_{2\alpha}$ and PGEs on cAMP production and intracellular calcium mobilization in avian uterine smooth muscle cells. Biol. Reprod. 36, 384–391.

Munsick, R.A., Sawyer, W.H., & van Dyke, H.B. (1960). Avian neurohypophyseal hormones: Pharmacological properties and tentative identification. Endocrinology 66, 860–871.

Nalbandov, A.V. (1976). *Reproductive Physiology of Mammals and Birds*, 3rd ed. Freeman, San Francisco.

Ohmichi, M., Koike, K., Nohara, A., Kanda, Y., Sakamoto, Y., Zhang, Z.X., Hirota, K., & Miyake, A. (1995). Oxytocin stimulates mitogen-activated protein kinase activity in cultured human puerperal uterine myometrial cells. Endocrinology 136, 2082–2087.

Olson, D.M., & Hertelendy, F. (1983). Avian shell gland contractility: Interaction of $PGF_{2\alpha}$ and arginine vasotocin with Ca^{2+}. Am. J. Physiol. 244, C150–C157.

Olson, D.M., Biellier, H.V., & Hertelendy, F. (1978). Shell gland responsiveness to prostaglandins $F_{2\alpha}$ and E_1 and to arginine vasotocin during the laying cycle of the domestic hen *(Gallus domesticus)*. Gen. Comp. Endocrinol. 36, 559–565.

Ralph, C.L., & Fraps, R.M. (1960). Induction of ovulation in the hen by injection of progesterone in the brain. Endocrinology 66, 269–272.

Riddle, O. (1921). A simple method of obtaining premature eggs from birds. Science 54, 664–666.

Rothchild, I., & Fraps, R.M. (1944). On the function of the ruptured ovarian follicle of the domestic fowl. Proc. Soc. Exp. Biol. Med. 56, 79–82.

Rzasa, J., & Ewy, Z. (1970). Effect of vasotocin and oxytocin on oviposition in the hen. J. Reprod. Fert. 21, 549–550.

Sharp, P.J. (Ed.). (1993). *Avian Endocrinology*. J. Endocrinol. Ltd., Bristol.

Shimada, K., & Saito, N. (1989). Control of oviposition in poultry. Crit. Rev. Poultry Biol. 2, 235–253.

Shimada, K., Olson, D.M., & Etches, R.J. (1984). Follicular and uterine prostaglandin levels in relation to uterine contraction and the first ovulation of a sequence in the hen. Biol. Reprod. 31, 76–82.

Shimada, K., Saito, N., Itogawa, K., & Koike, T.I. (1987). Changes in the plasma concentration of arginine vasotocin and prostaglandin F after intrauterine injection of prostaglandin $F_{2\alpha}$ and acetylcholine at various times during oviposition cycle of the domestic fowl *(Gallus domesticus)*. J. Reprod. Fert. 80, 143–150.

Tanaka, K., Goto, K., Yoshioka, T., Terao, T., & Koga, O. (1984). Changes in the plasma concentration of immunoreactive arginine vasotocin during oviposition in the domestic fowl. Brit. Poultry Sci. 25, 589–595.

Tóth, M., Olson, D.M., & Hertelendy, F. (1979). Binding of prostaglandin $F_{2\alpha}$ to membranes of shell gland muscle of laying hens: Correlation with contractile activity. Biol. Reprod. 20, 390–398.

Tóth, M., Asbóth, G., & Hertelendy, F. (1981). Selective binding of prostaglandin $F_{2\alpha}$ to membrane fractions enriched in 5′-nucleotides, Ca^{2+}-ATPase, and Mg^{2+}-(Na^+ +K^+)ATPase. Endocrinology 109, 106–112.

Tóth, M., Asbóth, G., & Hertelendy, F. (1983). Turnover of lipid-bound arachidonate and biosynthesis of prostanoids in the endometrium and myometrium of the laying hen. Arch. Biochem. Biophys. 226, 27–36.

Tóth, M., Gimes, G., & Hertelendy, F. (1987). Triton X-100 promotes the accumulation of phosphatidic acid and inhibits the synthesis of phosphatidylcholine in human decidua and chorion frondosum tissue *in vitro*. Biochem. Biophys. Acta 921, 417–425.

INITIATION OF PARTURITION IN NON-HUMAN PRIMATES

Jonathan J. Hirst and Geoffrey D. Thorburn

Advances in Organ Biology
Volume 1, pages 61–87.

ISBN: 1-55938-639-8

ABSTRACT

The endocrine regulation of pregnancy and the timing of labor onset differs markedly between species. Non-human primates are the closest experimental animals to humans and, therefore, studies of these species are of great value in bridging gaps in knowledge created by this divergence. The principal advantage of non-human primates for the study of pregnancy is the ability to perform maternal and fetal repetitive sampling protocols. Studies with chronically catheterized fetal rhesus monkeys have shown that labor begins at normal term following the death of the fetus, although the spread of delivery times around normal term is increased. These observations indicate that the fetus does not influence mean gestational length, but has a permissive role in the fine-tuning of the timing of labor onset. There is no decline in maternal progesterone concentration prior to labor onset in primates, a finding which is in marked contrast to other species. Since there is no fall in the sensitivity of the intrauterine tissues to progesterone, primate intrauterine tissues may be sufficiently refractory to progesterone to allow the initiation of labor in the presence of high gestational progesterone concentrations. There is a marginal increase in the concentration of estrogens as term approaches such that the ratio of estrogens to progesterone rises slightly. This rise may be responsible for the increase in gap junctions and oxytocin receptors that is observed prior to labor onset. Non-human primate studies have provided convincing evidence supporting a role for prostaglandins as the primary stimulatory factor in the initiation of labor. Longitudinal studies in the rhesus monkey have demonstrated that there is a steady rise in prostaglandin concentrations in the amniotic fluid before the onset of labor. These observations suggest that rising prostaglandin concentrations in the intrauterine tissues and amniotic fluid ultimately lead to the initiation of labor. Increasing intrauterine prostaglandin synthesis may be exclusively responsible for initiation of labor following the death of the fetus. Increased maternal oxytocin secretion most likely has a role in the stimulation of decidual prostaglandin synthesis at the time of spontaneous labor onset, whereas an increase in cytokine levels within the uterine compartment may be responsible for the stimulation of prostaglandin production following intrauterine infection. Pregnant rhesus monkeys exhibit episodes of markedly increased uterine activity at night which are driven by elevated oxytocin concentrations in the maternal plasma. The magnitude of the nocturnal rise in oxytocin concentrations increases as term approaches and leads to greater levels of nocturnal uterine activity. The development of nocturnal uterine activity episodes requires the presence of a suitably stimulated level of myometrial oxytocin receptors. This stimulation, in turn, is generated by the increasingly estrogenic environment produced by the fetus near term and may represent the contribution of the fetus to the timing of labor onset. The heightened level of nocturnal oxytocin secretion in the presence of elevated oxytocin receptor concentrations leads to a further increase in nocturnal uterine activity episodes near term. Once adequate levels of prostaglandins are attained within the intrauterine tissues, oxytocin-driven nocturnal uterine activity progresses to labor and delivery in non-human primate species.

I. INTRODUCTION

The regulation of pregnancy and the mechanisms involved in the initiation of parturition show numerous differences between species. Non-human primates are the closest experimental animals to humans. Therefore, studies of the mechanisms of the initiation of labor in non-human primate species are of great value for bridging the gaps in our knowledge that are created by this divergence. Important differences in the regulations of pregnancy between humans and non-human primates as well as within non-human primate species do exist and should not be overlooked when interpreting experimental results.

The major disadvantage of working with non-human primates is their high cost. In order to overcome this problem, non-human primate studies are largely designed to test findings in other species rather than for preliminary investigations. Therefore, studies proceed from either of two starting points; most obviously, the reinvestigation of findings made in lower species in a primate model. Secondly, for the study of clinical observation made in human pregnancy and for which ethical consideration prevents the inclusion of controlled experimental designs. A major advantage of the use of non-human primate species is the availability of repetitive sampling such that longitudinal studies of pregnancy may be performed. Table 1 sets out some of the advantages and disadvantages of working with non-human primates.

Studies of pregnancy and parturition have mainly been confined to a few species with the most notable work having been done in rhesus monkeys (*Macaca mulatta*), baboons (*Papio papio*), and marmosets (*Callithrix jacchus*). The larger size of the former two species has allowed the development of fetal and maternal chronically catheterized primate models for use in the determination of the fetal endocrine changes during pregnancy. Figure 1 illustrates a typical chronically catheterized rhesus monkey preparation. The jacket and tether system

Table 1. Advantages and Disadvantages of Non-human Primates in the Study of Pregnancy and Parturition

ADVANTAGES	1. Similarities to human pregnancy
	Singleton pregnancy
	Similar endocrinology
	Similar placentation
	Large size
	2. Experimental
	Longitudinal sampling possible
	Fetal and maternal sampling
DISADVANTAGES	High cost
	Dedicated facilities required

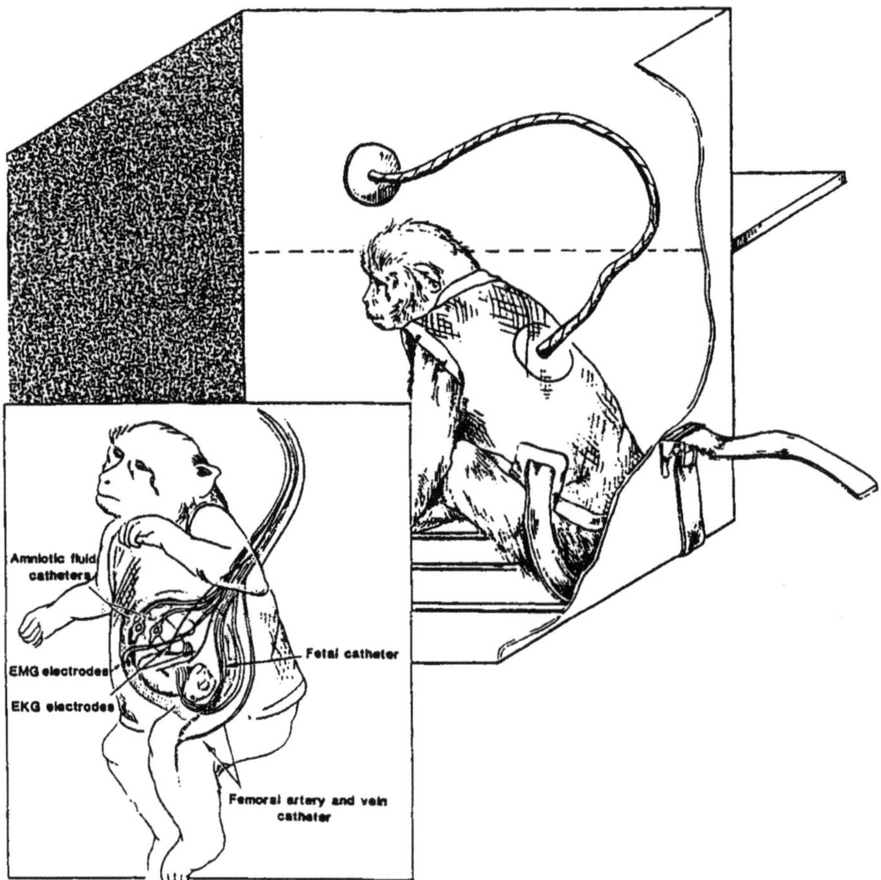

Figure 1. Diagrammatic representation of a typical fetal and maternal chronically catheterized rhesus monkey preparation. The jacket and tether system allow for fetal and maternal blood as well as amniotic fluid sampling while the animal is conscious and unrestrained.

allows for fetal and maternal blood as well as amniotic fluid sampling while the animal is conscious and unrestrained.

II. FETAL ROLE IN THE REGULATION OF THE TIMING OF LABOR ONSET

There are major differences in the mechanisms that appear to be responsible for triggering the onset of labor between mammalian species (Liggins and Thorburn, 1994). Many of these differences reflect variations in the contribution of fetal factors to the control of the initiation of labor. The available evidence suggests that the primate fetus has no more than a permissive role in the control of the mechanisms initiating parturition. Neither fetal death in the absence of infection, nor the surgical removal of the rhesus monkey fetus influences the mean length of gestation (Novy et al., 1977). The spread of delivery times around the mean values was increased, however, suggesting that fetal factors have a role in fine-tuning the timing of labor onset.

The continuous infusion of glucocorticoids does not lead to the induction of premature parturition in rhesus monkeys (Novy et al., 1977). This finding indicates that the fetal hypothalamo-hypophyseal-adrenal axis is not involved in controlling the length of gestation. This is similar to human pregnancy in which anencephalic fetuses are delivered at normal term. These observations represent a marked difference from the mechanisms that operate in lower mammalian species, where maturation of the fetal hypothalamo-hypophyseal-adrenal axis leads to an increase in glucocorticoid production near term and this initiates processes that results in the onset of labor. In primates long-term exposure to exogenous glucocorticoids and the consequent suppression of fetal adrenal function result in a reduction of adrenal androgen production and a fall in maternal estrogen concentrations (see following). The death of the fetus also leads to a loss of androgens available for estrogen production. This decline in estrogen levels is likely to be responsible for the increased spread of deliveries around the mean gestational length that is observed following fetal death or glucocorticoid infusions in rhesus monkeys. The finding that maternal hypophysectomy, which would also suppress estrogen levels somewhat, is also associated with parturition at normal term lends support to the view that mean gestational length is determined by intrauterine rather that maternal endocrine changes (Chez et al., 1970).

There appear to be multiple levels in the regulation of the timing of labor onset in rhesus monkeys. Intrauterine factors control mean gestational length, whereas the precise day of delivery is influenced by fetal mechanisms involving the modulation of estrogen precursor production (see following). In rhesus monkeys labor and delivery occur exclusively at night with labor commencing during the early hours of darkness (Hirst et al., 1993a). This observation indicates that the exact hour at which labor commences is regulated in this species. Since the timing of labor may be slowly altered by changing the light:dark environment, maternal factors (discussed in the following section) determine the exact time during the day that labor is initiated. Therefore, although the fetus contributes to the precision of the timing of delivery, maternal influences predominate in the final triggering of the initiation of labor in this species.

III. PROGESTERONE

Progesterone is secreted throughout pregnancy by the placenta and corpus luteum in primate species. The contribution of the placenta predominates by mid-gestation since removal of the placenta by cesarean section leads to a marked fall in progesterone concentrations (Walsh et al., 1977). During early pregnancy in rhesus monkeys there is a steady rise in progesterone concentrations which reach a plateau by mid-gestation. Concentrations remain at these levels throughout late gestation. Although not dramatic, progesterone concentrations increase further from five days before term and continue to rise until the delivery (Novy and Walsh, 1983). This is very similar to the steady or slightly increased levels of progesterone that are observed at the time of labor onset in women. There is a marked fall in progesterone concentrations following delivery of the placenta, however, there is no decline in progesterone concentrations prior to the onset of labor in rhesus monkeys. This is in marked contrast, however, to the fall in progesterone levels that is observed before the onset of labor in lower mammalian species.

Diurnal rhythms in both fetal and maternal progesterone concentrations are observed from as early as day 30 of pregnancy in rhesus monkeys (term 167 days; Challis et al., 1980). By late gestation maximal levels of progesterone are observed at night and show an approximately inverse pattern to that of estrogens and cortisol levels. Although the reasons for the variation in progesterone levels are unclear, Hess et al. (1981) have suggested that the pattern in progesterone levels may be a reflection of the cortisol pattern. This is possibly brought about by the displacement of progesterone from binding proteins by cortisol. Therefore, the variation in progesterone levels may be secondary to changes in cortisol levels and fetal adrenal activity.

Estrogen concentrations rise near term and are more marked compared to that of progesterone. This elevation in estrogen concentrations leads to a small increase in the ratio of estrogens to progesterone which may result in a lessening of the influence of progesterone. However, this change is not of sufficient magnitude to bring about alteration in the concentrations of uterine steroid hormone receptors which are predominantly regulated by circulating estrogen levels. Wilson et al. (1991) have measured diurnal changes in estrogen and progesterone concentrations at the time of labor in baboons. These investigators found that estrogen levels were maximal in the early evening before progesterone levels had started to rise. This pattern of secretion results in a window of relatively low progesterone levels in the early evening. These authors proposed that this window, possibly mediated by a rise in myometrial oxytocin receptors and gap junctions, allows the initiation of labor at this time. The formation of gap junctions and oxytocin receptors is susceptible to inhibition by progesterone. However, the finding that injections of progesterone do not prolong labor in primate species is not consistent with this proposal.

The concentrations of gap junctions and oxytocin receptors (see following) rise during late gestation in rhesus monkeys and a rise in the ratio of estrogen to progesterone may be involved in this increase (Haluska and Novy, 1993). However, the inability of high levels of progesterone to inhibit labor in rhesus monkeys suggests an adequate increase in gap junctions and oxytocin receptors can develop in the presence of elevated progesterone concentrations.

RU486 is a potent 19-norsteroid synthetic progesterone receptor antagonist. Administration of RU486 antagonizes progesterone and results in the induction of labor at any time during gestation in women as well as rhesus monkeys. Haluska et al. (1987) used RU486 to induce labor during late pregnancy in rhesus monkeys while monitoring uterine activity. These investigators found that although myometrial activity was increased and labor eventually resulted from RU486 treatment, the myometrial activity displayed a contractile pattern that was markedly different to the pattern seen at the onset of normal labor. There was also a delay in the increase in amniotic fluid prostaglandin concentrations that is seen around the time of normal labor. Furthermore, cervical dilatation did not occur following RU486 treatment until uterine activity reached very high levels. In addition, gap junction levels in myometrial tissue, obtained after RU486 treatment, were elevated to levels markedly beyond those seen after normal labor. These findings indicated that while effective in stimulating uterine activity, withdrawal of progesterone by RU486 treatment does not produce the orderly myometrial activity and cervical changes seen at normal labor onset.

The above studies of progesterone withdrawal, using RU486, indicate that progesterone does have a suppressive action on the myometrium. The finding that progesterone causes hyperpolarization of myometrial muscle cells in isolated human myometrial strip preparations supports a suppressive role for progesterone (Garfield, 1988). The finding that labor will proceed successfully despite gestational plasma progesterone levels suggests that the primate uterus may be, to some extent, insensitive to the suppressive actions of progesterone. Romero et al. (1988) have reported that amniotic fluid estradiol, estriol, and dihydroepiandrosterone sulfate (DHEAS) concentrations increase with labor onset in women. No change in progesterone levels was observed and the ratio of estrogens to progesterone was consequently increased. They suggested that a local increase in this ratio in the amniotic fluid may be a factor in the initiation of labor. In rhesus monkeys, amniotic estrogen levels also increase toward the time of labor onset (Novy and Walsh, 1983). However, stimulation of a premature rise in amniotic estrogen levels, with ACTH infusions, does not induce preterm labor. In addition, normal labor is not induced by insertion of estrogen containing implants into the amniotic cavity (Novy and Haluska, 1988). These studies do not support a local change in the estrogen to progesterone ratio in the control of the onset of labor.

The possibility that the myometrium becomes insensitive to the suppressive effects of progesterone or that estrogen sensitivity increases prior to the onset of labor has been investigated by examining myometrial steroid hormone receptor concentrations in rhesus monkeys. Haluska et al. (1990) measured estrogen receptor concentrations in myometrium and found that receptor concentrations did not increase before normal labor onset, but were increased by RU486 treatment. These authors concluded that, although progesterone did suppress estrogen receptor concentrations during pregnancy, an increase in receptor concentrations was not necessary for the initiation of labor. These findings, therefore, indicate that an increase in these receptors is not a process that contributes to initiation of labor. We have also investigated progesterone receptor concentrations before and after labor onset in rhesus monkeys. We found that there was no change in the concentration of progesterone receptors in either decidual or myometrial tissue with labor onset (Hirst et al., 1990). There was also no redistribution of receptors between the nucleus and cytoplasm in these tissues. These studies indicate that there is no lessening of receptors available for progesterone binding at term and suggest that there is no loss of decidual or myometrial sensitivity to progesterone at the time of labor onset. Together, these findings indicate that neither a local decline in progesterone concentration nor a change in the sensitivity to progesterone occurs prior to labor and, therefore, parturition takes place despite the presence of the suppressive effects of progesterone.

IV. ESTROGENS

The mechanisms involved in estrogen production differ markedly between primates and lower mammalian species. The predominant estrogen during pregnancy also differs within primate species. Estrone is the major estrogen in the fetal circulation of the rhesus monkey, whereas 17β-estradiol predominates in the maternal circulation (Resko et al., 1975). The fetal rhesus monkey has a functional fetal-placental unit in which the fetus contributes precursors for placental estrogen production. In lower primates, as in human pregnancy, placental estrogen production is at least partially dependent on androgens produced by the fetal adrenal gland. The rhesus monkey placenta contains abundant sulfatase activity allowing the utilization of DHEAS as well as DHEA (dihydro-epiandrosterone) for aromatization. Fetal androgens destined for estrogen production originate in the fetal zone of the fetal adrenal glands. The fetal adrenal cortex of the rhesus monkey shows functional specialization similar to that seen with the human fetus. The inner fetal zone secretes DHEA and DHEAS, whereas the outer zone, which differentiates into the adult cortex, is responsible for fetal cortisol production. Both androgen production by the fetal zone and cortisol secretion are stimulated by fetal ACTH (Liggins and Thorburn, 1994). The development of the fetal hypothalamo-hypophyseal-adrenal axis with advancing gestation leads to increased androgen secretion by the fetal adrenal glands. This, in turn, is responsible for the rising levels of estrogens during late gestation and toward term. Compared to the human fetus, the fetal rhesus monkey contributes a considerably lower proportion of total androgens for estrogen production. The death of the rhesus monkey fetus during late gestation leads to a decline in maternal estrogen levels of between

25–50% (Novy et al., 1977). The remaining androgens utilized for estrogen synthesis are derived from the maternal adrenal glands.

Multiple blood sampling from chronically catheterized rhesus monkeys has allowed the determination of the diurnal pattern of changes in the concentration of estrogens in this species. Maternal estrogen concentrations show a marked diurnal rhythm during late gestation. Maximum concentrations are observed around 2400 h (Walsh et al., 1988). This diurnal pattern is very similar to the diurnal pattern in DHEA levels observed at this time. Fetal cortisol concentrations also show a diurnal rhythm, but this rhythm displays an approximately inverse relationship to the pattern of estrogen concentrations with minimum concentrations observed during the light phase of the photoperiod 0700–2100 h (Walsh et al., 1988). These findings indicate that the rhythm in estrogen levels may result from the diurnal changes in the levels of fetal androgens. Furthermore, the pattern in fetal DHEA and DHEAS concentrations may be driven by the inhibitory influence of cortisol on fetal ACTH secretion. The factors responsible for initiating the diurnal variation in fetal and maternal cortisol levels remain to be determined. However, the diurnal pattern is shifted by altering the photoperiod suggesting that the maternal circadian system may play a central role.

Novy et al. (1977) have found that, in rhesus monkeys, death of the fetus lowers maternal plasma estrogen concentrations, but does not alter the mean length of gestation. Similarly, in baboons, removal of the fetus, by surgical fetectomy, results in delivery of the placenta at approximately normal term despite markedly lower estrogen concentrations. The time of delivery, however, showed a large spread around normal term. In addition, some rhesus monkeys showed a prolongation of gestation after fetal death. In such animals, estradiol infusions rapidly induced labor onset (Novy et al., 1977). These findings suggest that, although mean gestational length may be determined by other factors, the rise in estrogen concentrations at term may provide the final stimulus for the onset of the initiation process.

In rhesus monkeys estrogen receptors are present throughout gestation in the myometrium and decidua, but at very much lower levels than those observed during the menstrual cycle (Haluska et al., 1990). The low receptor levels most likely result from the long period of suppression by the elevated progesterone levels during gestation. There is no increase in estrogen receptor levels in either the decidua or myometrium at term. This is consistent with the maintenance of suppression by the steady or slightly increased progesterone levels around the time of labor onset. The low levels of estrogen receptors at term may explain why estrogens exert relatively little influence ovėr the timing of delivery in primate species.

V. ROLE OF PROSTAGLANDINS

The role of prostaglandins in the initiation of labor in women has been discussed extensively elsewhere and it is only necessary to discuss studies involving non-human primates in this chapter. Non-human primate studies, however, have produced several important findings that are sufficiently noteworthy to warrant inclusion. Furthermore, the ability to repetitively obtain amniotic fluid samples from a non-human primate during late gestation has allowed the accurate determination of prostaglandin concentrations in the amniotic fluid at term. This has contributed supporting evidence for a triggering role for local changes in prostaglandin concentrations in the initiation of labor.

Evidence supporting a causative role for prostaglandins in the initiation of labor in rhesus monkeys is similar to that for human pregnancy, but more conclusive. Prostaglandin E_2 (PGE_2) and prostaglandin $F_{2\alpha}$ ($PGF_{2\alpha}$) administration induce preterm labor in rhesus monkeys, whereas, the inhibition of prostaglandin synthesis by indomethacin treatment prolongs pregnancy (Novy et al., 1974). Mitchell et al. (1986) have measured $PGF_{2\alpha}$ and the metabolite of $PGF_{2\alpha}$, PGFM, in the maternal plasma and amniotic fluid of rhesus monkeys. These authors did not find changes in the concentration of either prostaglandin in the plasma at term. In the amniotic fluid, however, $PGF_{2\alpha}$ and PGFM levels increased markedly during the five days preceding labor. Walsh (1991) further investigated amniotic $PGF_{2\alpha}$ concentrations in catheterized rhesus monkeys during late gestation. These investigators observed a marked prepartum increase in amniotic $PGF_{2\alpha}$ concentrations in some animals and small increases in a subgroup of the animals studies. The animals that showed a substantial increase in $PGF_{2\alpha}$ preceding labor onset also developed strong nocturnal uterine activity episodes on the nights preceding labor. Since PGE_2 levels were not measured in these studies, it is possible that those animals that showed a lesser rise in $PGF_{2\alpha}$ levels developed higher levels of PGE_2. This may have led to more rapid cervical dilation so that labor and delivery was possible with a smaller increase in $PGF_{2\alpha}$ levels. Haluska et al. (1987) in a previous study with rhesus monkeys, reported that mean amniotic fluid $PGF_{2\alpha}$, PGFM, and PGEM (the major metabolite of PGE_2) concentrations increased significantly before the onset of labor. These authors showed that PGEM concentrations rose to the greatest extent. They also reported that there was an increase in all these prostaglandins during late gestation immediately prior to the rise in uterine activity that followed RU486 treatment. These findings suggest that the rise in uterine activity induced by RU486 was driven by increased prostaglandin production by the intrauterine tissues. Furthermore, these findings suggest that an increase in intrauterine prostaglandin production is a primary event in the initiation of uterine activity and the onset of normal labor in rhesus monkeys.

In addition to prostaglandins, Haluska et al. (1990) and Walsh (1991) found, in two separate studies with catheterized rhesus monkeys, that amniotic fluid concentrations of the lipoxygenase products LTB_4, LTC_4, and 5-HETE (leukotrienes B_4, C_4, and 5-hydroxy-eicosatetraenoic acid, respectively) were elevated prior to the onset of labor. Although LTC_4 has little oxytocic activity, LTC_4 and 5-HETE are potent stimulators of human myometrial strips. These findings suggest that these lipoxygenase products may contribute to the initiation of labor-associated uterine activity in rhesus monkeys.

The elevated levels of prostaglandins in the amniotic fluid prior to labor onset suggest that the production of prostaglandins by prostaglandin synthase enzymes is induced during the process of labor initiation in intrauterine tissues. The finding that the injection of arachidonic acid into the amniotic fluid does not initiate labor in rhesus monkeys, but that amniotic injections of $PGF_{2\alpha}$ induces preterm labor onset suggests that precursor availability is not a limiting factor in prostaglandin production at term (Robinson et al., 1978). Together, these observations suggest that the activity of prostaglandin synthase is the limiting step in the production of prostaglandin at term in rhesus monkeys and the factor(s) that regulate enzyme activity are pivotal to the initiation of labor.

The factors that regulate prostaglandin synthase activity at the onset of normal labor have not been investigated in non-human primates. However, studies with non-human primate species have contributed greatly to our understanding of the mechanisms by which intrauterine infection causes preterm labor. These mechanisms may involve the induction of prostaglandin synthase enzymes. Gravett et al. (1994) have recently described the sequence of events following bacterial invasion of the chorio-amnion and the amniotic cavity in a rhesus monkey model. First, these authors showed that the cervical-vaginal flora in healthy rhesus monkeys was similar to that observed during human pregnancy. Importantly, the potentially pathological bacteria implicated in human intra-amniotic infection were also isolated from rhesus monkeys. These investigators then studied infection by introducing bacteria into the amniotic fluid of chronically catheterized animals. In these studies group B streptococci were inoculated into the amniotic fluid and uterine activity was monitored. Following bacterial growth, there was a marked increase in uterine activity and a rise in amniotic fluid $PGF_{2\alpha}$ and PGE_2. The rise in prostaglandin levels preceded labor onset which occurred within three days of treatment in all animals. Gravett et al. (1994) also showed that amniotic fluid concentrations of the inflammatory cytokines interleukin-1β (IL-1β), interleukin-6 (IL-6), and tumor necrosis factor-α (TNF-α) were elevated following infection and before the onset of labor. These authors found that an increase in these cytokines in response to infection was required to initiate preterm labor. The presence of bacteria alone was not sufficient to increase prostaglandin levels in the amniotic fluid or to induce preterm labor. The finding that amniotic fluid concentrations of PGE_2 was elevated to a greater extent than $PGF_{2\alpha}$ following amniotic fluid infection is consistent with preferential production of PGE_2 by amnion cells. In human pregnancy $PGF_{2\alpha}$ is produced mainly by the decidua. This suggests that ascending infection which reaches the chorio-decidua may stimulate preterm labor by increasing $PGF_{2\alpha}$ production and uterine activity. If infection reaches the amniotic fluid, the resultant increase in PGE_2 may lead to delivery primarily as a result of loss of cervical competence.

Using the rhesus monkey model, Gravett et al. (1994) have determined the role of cytokine release in the initiation of labor in response to infection. These investigators found that graded infusions of IL-1β into the amniotic fluid led to a dose-dependent increase in $PGF_{2\alpha}$ and PGE_2 production. Furthermore, the pattern of the increase closely resembled that observed following the introduction of bacteria into the amniotic fluid. The rise in prostaglandin levels was followed by an increase in uterine activity. After cessation of the infusions prostaglandin concentrations returned to basal levels and uterine activity subsided. These findings suggest that infection induced preterm labor involves the ordered release of cytokines, with IL-1β being produced after the onset of the inflammatory response. IL-1β stimulates prostaglandin synthesis by the intrauterine tissues, followed by the induction of uterine activity and cervical softening. The use of a non-human primate model for these studies has resulted in the stepwise order of events in the induction of preterm labor in the presence of infection to be determined. The elucidation of these steps will be useful for the early detection of infection-induced preterm labor as well as providing a point where preventative treatments may be initiated.

VI. RELAXIN

Relaxin has been detected in the plasma of rhesus monkeys from early pregnancy onward (Weiss et al., 1981). Concentrations of relaxin are maintained at levels similar to those observed during early pregnancy throughout gestation. Relaxin depresses oxytocin-induced contractions by isolated human myometrium. There is no decline in relaxin levels prior to labor onset in rhesus monkeys; therefore, it is unlikely that a fall in circulating relaxin concentration has a role in increasing myometrial activity at term (Weiss et al., 1981). Infusions of human relaxin do not inhibit uterine activity episodes in rhesus monkeys. This may be because the concentrations of this highly labile peptide were not sufficient or because of a lack of affinity of human relaxin for the rhesus monkey relaxin receptor (Dr. G. J. Haluska, personal communication). Although these possibilities have not been investigated, the overall findings do not lend support for a role for circulating relaxin in the regulation of myometrial activity during late gestation. Relaxin is produced primarily by the corpus luteum during pregnancy and this site of production is responsible for circulating relaxin. Relaxin has been detected in intrauterine tissues suggesting local synthesis, as has been reported for human intrauterine tissues. The possibility of a role for this relaxin in the control of uterine activity and cervical dilatation at term has not been investigated in non-human primates.

VII. OXYTOCIN AND UTERINE ACTIVITY DURING LATE PREGNANCY

Despite numerous studies, an endocrine role for oxytocin in the initiation of labor is controversial. The most compelling evidence supporting a role for oxytocin in labor initiation is the observation that exogenous oxytocin, administered near term, induces uterine activity indistinguishable from that observed with spontaneous labor onset. The finding that various agents that inhibit oxytocin secretion reduce the frequency and amplitude of uterine contractions also supports a major role for oxytocin in labor. The inability, however, to find a consistent rise in maternal oxytocin concentrations prior to the time of labor onset in women has led to uncertainty over the role of oxytocin in the initiation process.

Several studies in non-human primates have led to a better understanding of the involvement of oxytocin in primate labor (Hirst et al., 1993a). The use of repetitive blood sampling protocols is of critical importance in the measurement of oxytocin, since circulating levels of the peptide change rapidly. Furthermore, because uterine activity can be monitored continuously in non-human primates, oxytocin levels may be correlated with uterine activity measurements.

Pregnant rhesus monkeys display 24 h rhythms in uterine activity with episodes of increased uterine activity during the hours of darkness (Ducsay et al., 1983). These episodes begin in the early evening and are of variable magnitude, but maximal uterine activity is usually observed around midnight (Hirst et al., 1991). We investigated the role of oxytocin in the generation of these episodes of uterine activity. In our laboratory pregnant rhesus monkeys were catheterized between 121–130 days of gestation to allow oxytocin measurement and intra-amniotic fluid pressure recording. Figure 2a shows an intrauterine pressure recording of a typical nocturnal uterine activity episode. We found that infusion of an oxytocin antagonist (ORF 22164, Atosiban) during uterine activity episodes resulted in the inhibition of further activity (Hirst et al., 1991). Figure 2b shows the effect of the antagonist on an episode of uterine activity at 141 days of gestation for comparison with Figure 2a. As shown in Figure 2b, the episode of uterine activity developed around the start of the dark phase of the photoperiod (2300 h) and increased in strength up to 2400 h. During the following 90-minute oxytocin antagonist infusion uterine activity was markedly reduced to baseline levels. After termination of the infusion, activity increased to levels similar to preinfusion values.

The occurrence of nocturnal uterine activity varied between animals in the 130–160 day range of gestational ages with some animals showing nocturnal episodes of uterine activity throughout this period. Alternatively, some animals showed little nocturnal activity until 10–15 days prior to term. All animals, however, showed nocturnal uterine activity as term approached. Our results with the oxytocin antagonist indicated that these episodes were either driven by rising plasma oxytocin concentrations or by changing sensitivity to basal levels of oxytocin. To investigate these possibilities we measured maternal plasma oxytocin at 3 h intervals over 24 h in animals that displayed significantly elevated levels of uterine activity at night. Figure 3 shows maternal oxytocin levels and uterine activity over the dark phase of the photoperiod. There was a significant rise in oxytocin concentrations at night and this rise was correlated with elevated uterine activity. We have also found, using a 15-minute sampling interval, that the rise in oxytocin levels precedes the onset of uterine activity at night (Hirst et al., 1991). These data show that nocturnal uterine activity during late gestation is primarily driven by increased oxytocin levels in the maternal plasma.

Nocturnal uterine activity episodes occur more consistently as gestation advances with all animals showing increased uterine activity at night

A

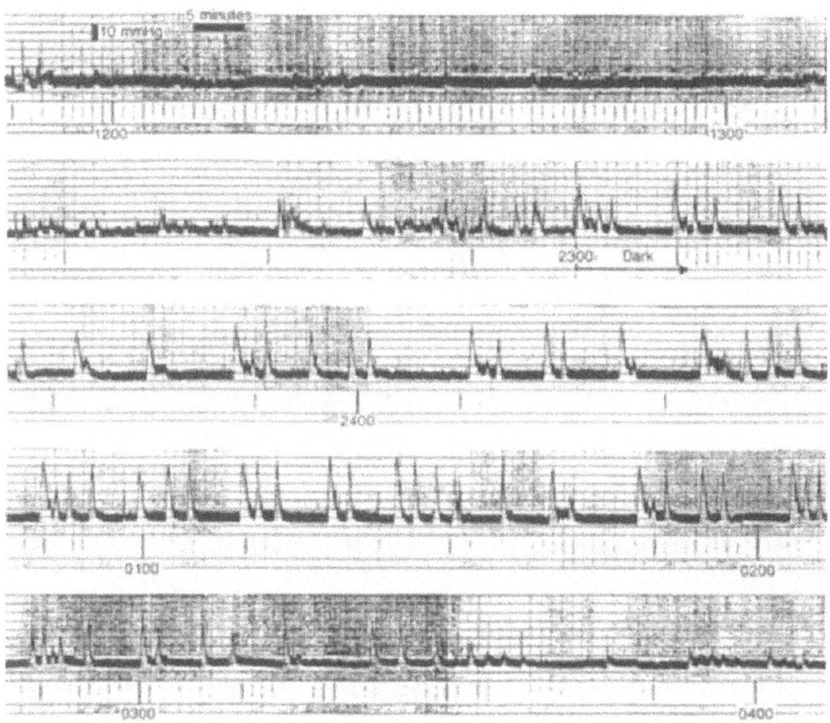

Figure 2. (a) Intra-amniotic pressure recording showing an episode of nocturnal uterine activity. Uterine activity starts around the beginning of the dark phase, reaches maximal levels around 2400 h, and declines in the later hours of the dark phase of the photoperiod (0300–0400 h). (b) Intra-amniotic pressure recording showing an episode of nocturnal uterine activity and the effect of an oxytocin antagonist (ORF 22164, Atosiban) infusion. Uterine activity increased around the beginning of the dark phase of the photoperiod (2300 h) and the oxytocin antagonist infusion was started 60 minutes after the start of the dark phase when uterine activity was established. Both the frequency and amplitude of contractions were reduced during the 90 minute antagonist infusion. (Modified from Hirst et al., 1993a, with permission.) (Continued.)

B

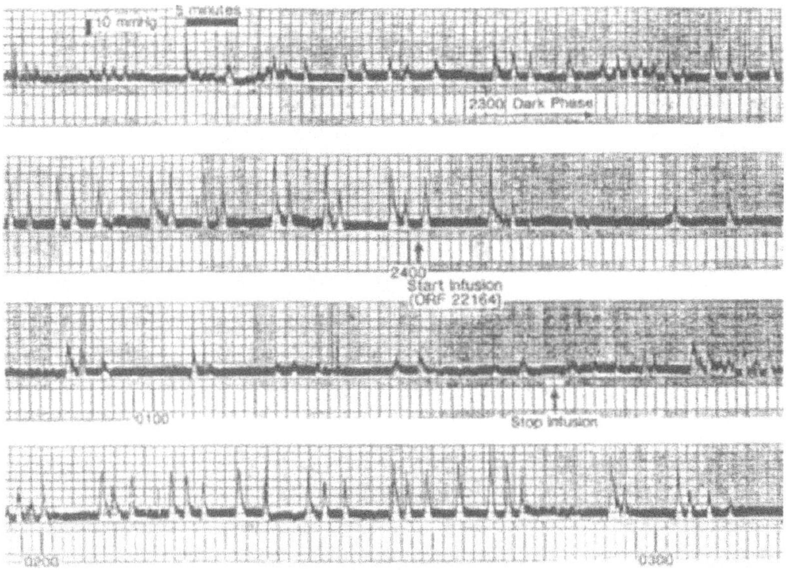

Figure 2. Continued

near term. In addition, the magnitude and duration of the nocturnal activity increases as term approaches and this activity is most pronounced on the nights immediately preceding spontaneous labor onset (Haluska et al., 1987). We have investigated the role of oxytocin in the rise in nocturnal uterine activity levels during late pregnancy by concurrently measuring oxytocin concentrations and uterine activity at weekly intervals during late gestation (Hirst et al., 1993b). We found that there was a significant increase in nocturnal oxytocin levels with advancing gestation and this increase was correlated with the increasing levels of nocturnal uterine activity as term approached. Our findings suggest that the increase in nocturnal uterine activity as term approaches results from rising levels of oxytocin in the maternal plasma and this rise begins many days prior to delivery. Interestingly, despite the rise in nocturnal oxytocin levels, there was no change in levels measured during the day. Therefore, studies which do not incorporate nighttime sampling would not detect an upward progression in oxytocin levels toward term. Nighttime delivery is typical for rhesus monkeys and our findings suggest that elevated oxytocin levels and nocturnal uterine activity ultimately lead to the initiation of labor at night in this species.

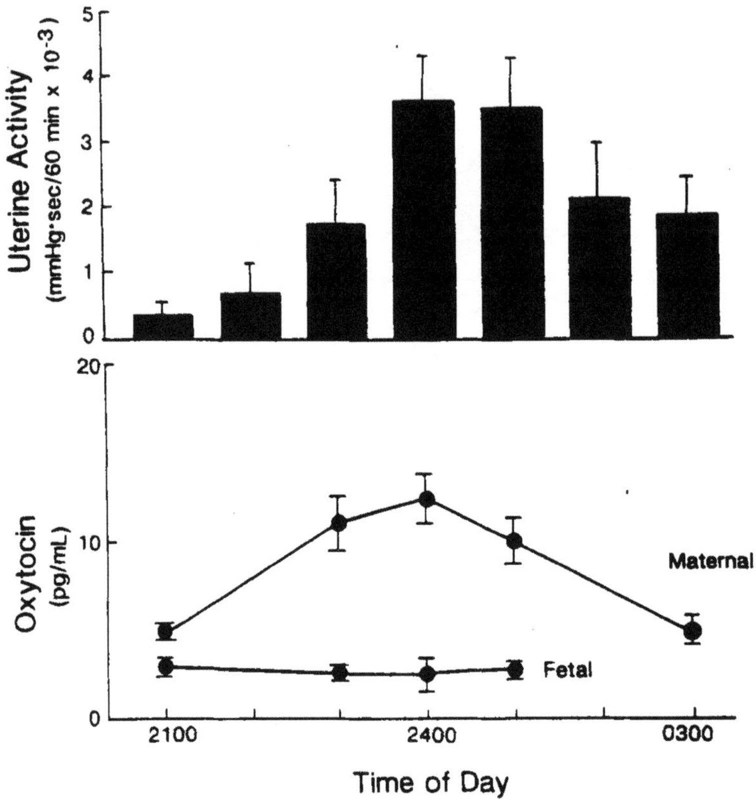

Figure 3. Uterine activity and oxytocin concentrations in the maternal and fetal plasma measured during the night ($n = 4$). Oxytocin concentration was significantly elevated in the maternal plasma during the hours of darkness (2100–0100 h, $p < 0.05$). The animals were maintained on a photoperiod with the dark phase between the dark phase 2300–0700 h. There was also no increase in oxytocin concentrations at any time in the fetal plasma. Vertical bars indicate the S.E. (Reproduced from Hirst et al., 1993b, with permission.)

VIII. OXYTOCIN CONCENTRATIONS AT TERM AND DURING LABOR

The role of oxytocin in the process of initiation of labor in women is unclear. However, considerable evidence has accumulated supporting a role for oxytocin in facilitating the expulsion of the fetus (see Hirst et al., 1993a for review). We have measured oxytocin levels on the night of delivery in rhesus monkeys. All monitoring and sampling were performed without disturbing the animals so that any possible effects of stress would be negated and all animals delivered during the night. Figure 4 shows oxytocin concentrations during the hours leading up to delivery of the fetus (time 0). Oxytocin levels were basal between 12 h and 5 h before delivery, but rose significantly 3 h before delivery. Both oxytocin and uterine activity increased further as labor progressed and were significantly correlated. The pattern of uterine activity and oxytocin levels between 3 h and 1 h before delivery were very similar to those seen during a nocturnal uterine activity episode near term. The results suggest that cervical changes may be the critical factor in the progression from a large nocturnal uterine activity episode to labor. As shown in Figure 4, oxytocin concentrations increase further during delivery (–30 minutes to 0). The very high concentrations of oxytocin attained during expulsion of the fetus may result from the additive effect of nocturnal oxytocin secretion and stimulation to further oxytocin secretion by the Ferguson reflex.

Studies in women have shown that oxytocin levels vary greatly during late pregnancy, with peaks of elevated concentrations on a background of much lower levels. This is consistent with pulsatile release from the neurohypophysis (Fuchs et al., 1991). The pulsatility in oxytocin concentrations appears to be much greater in women compared to the relatively steady levels we have observed in rhesus monkeys. The differences may result from more rapid degradation of oxytocin in women leading to large falls in concentrations between secretory pulses. This pulsatility may also account for some of the difficulties in measuring oxytocin levels consistently in human pregnancy.

The fetal neurohypophysis contains appreciable amounts of oxytocin during late gestation. Chard et al. (1971) have reported that human umbilical cord blood contains high levels of oxytocin and that this may contribute to maternal oxytocin levels as well as uterine activity at labor in women. Maternal oxytocin levels were lower compared to fetal levels suggesting that a suitable concentration gradient for transfer was present. However, the degradation of oxytocin by the abundant cystine aminopeptidase in the placenta would likely limit the passage of intact oxytocin. To investigate the possibility of fetal oxytocin contributing to maternal concentrations at term, we concurrently measured fetal and maternal oxytocin concentrations in rhesus monkeys during late gestation. As shown in Figure 3, in contrast to maternal plasma, we found there was no 24 h rhythm in fetal levels (Hirst et al., 1993b). The absence of a rise in fetal concentrations at night when maternal levels are elevated suggests that the fetus does not contribute to uterine activity or maternal

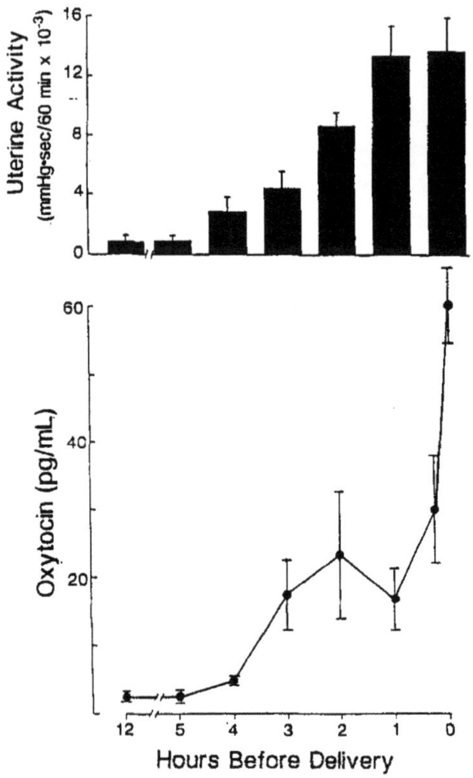

Figure 4. Uterine activity and maternal plasma oxytocin concentrations measured in the hours before delivery in six animals that delivered during the hours of darkness. Oxytocin concentrations and uterine activity were significantly elevated in the 1–3 hours before delivery compared to 4–12 hours before delivery. Uterine activity was also correlated with oxytocin concentrations ($r = 0.73$, $p < 0.03$). Vertical bars indicate the S.E. (Reproduced from Hirst et al., 1993b, with permission.)

concentrations at this time. Fetal oxytocin levels were basal throughout our studies. We also simultaneously measured fetal and maternal oxytocin levels at term in two monkeys and during labor in one monkey. Although maternal oxytocin levels were markedly elevated at night as well at 3 h and 1 h prior to delivery, fetal levels were either basal or undetectable at all times examined. These findings indicate that the fetus does not contribute to the elevated maternal oxytocin levels at labor in rhesus monkeys.

We have used graded fetal oxytocin infusions to further investigate if fetal oxytocin reaches the maternal circulation during pregnancy (Hirst et al., 1993b). Oxytocin was infused into the fetal jugular vein and fetal as well as maternal plasma oxytocin levels were measured. As expected there was a very large increase in fetal oxytocin concentrations during the infusions. There was, however, no increase in maternal oxytocin levels even though very large doses of oxytocin were infused and very high fetal oxytocin concentrations were reached. There were also no changes in uterine activity during the infusion or for at least 6 h after the infusions were discontinued. These results support our contention that, in a stable chronically catheterized animal, oxytocin does not cross the primate placenta. Our results, however, do not exclude the possibility that oxytocin escapes the fetal circulation during delivery of the fetus and may influence uterine activity at this time. Dawood et al. (1979) have reported that a bolus injection of oxytocin in fetal baboons leads to a rise in maternal oxytocin concentrations. These studies were performed at cesarean section, thus leaving open the possibility that the results were influenced by changes in the maternal-fetal vascular relationship at delivery.

IX. OXYTOCIN SENSITIVITY CHANGES DURING PREGNANCY

The sensitivity of the myometrium to oxytocin is maximal near term in all species studied so far (Fuchs et al., 1991). In women the sensitivity of the myometrium to oxytocin is significantly elevated by mid-pregnancy compared to the early pregnant and non- pregnant myometrium. We have found oxytocin sensitivity also increased progressively over the five nights preceding delivery in rhesus monkeys. As described above, we also found that nocturnal plasma oxytocin concentration rose with advancing gestation (Hirst et al., 1993b). There was no further increase in the magnitude of the nocturnal rise over the two nights before labor onset, although uterine activity did increase. These observations indicate that oxytocin sensitivity increases during this period and suggest that increasing oxytocin sensitivity as well as rising plasma levels contribute to the initiation of labor at night in rhesus monkeys (Hirst et al., 1993b).

Myometrial oxytocin receptors are of high oxytocin binding affinity with an equilibrium dissociation constant (Kd) of 1–3 nM and similar values have been observed in other tissues (Fuchs et al., 1982). As with sensitivity, myometrial oxytocin receptor concentrations are much higher in late pregnancy than in the non-pregnant myometrium. In women, at term, receptor concentrations are 100-fold greater than in nonpregnancy (Fuchs et al., 1982). These receptor concentration changes are consistent with the observed heightened levels of oxytocin sensitivity and support the view that the concentrations of circulating oxytocin need only rise by a small extent to elicit uterine contractions at term. These receptor changes are, therefore, also responsible for the low doses of exogenous oxytocin required to initiate contractions at term. We have found that oxytocin receptor concentrations rise, in parallel with oxytocin sensitivity, at the time of labor in rhesus monkeys (Hirst et al., 1991). These observations indicate that labor at night in rhesus monkeys may result from the combined effect of elevated nocturnal plasma oxytocin concentrations and increased oxytocin receptor levels.

The factors that regulate oxytocin receptor levels in primates remain unclear. Studies in rats, however, suggest the steroid hormone environment has a critical role. Oxytocin receptor concentrations in the rat show a pattern of changes during gestation that is similar to that observed in women, with receptor levels rising during late gestation followed by a sharp increase at term (Fuchs et al., 1983). Parturition in rats is preceded by a rise in estrogen levels and a fall in progesterone concentrations and these changes are correlated with rising receptor levels. Fuchs et al. (1983) have shown by treating rats with steroids that estrogen treatment up-regulates oxytocin receptors, whereas progesterone blocks this effect by estrogens. Estrogen treatment in women during late pregnancy does increase oxytocin sensitivity suggesting that estrogens stimulate oxytocin receptor concentrations and are, at least in part, responsible for the increase in receptor levels in late pregnancy (Pinto et al., 1964). Novy et al. (1977) have reported that the fall in maternal estradiol concentrations following fetal death results in the suppression of nocturnal uterine activity episodes in rhesus monkeys. These investigators also showed that the episodes were reestablished by estradiol infusions and suggested that the fall in estradiol concentration was responsible for the inhibition of nocturnal uterine activity. These results suggest that a critical level of estrogens is required to maintain oxytocin receptor concentrations so that the nocturnal rise in oxytocin concentrations can initiate uterine activity. These findings further indicate that the permissive role of estrogens in the initiation of labor in rhesus monkeys likely involves the maintenance of oxytocin receptor concentrations.

The role of progesterone in the regulation of receptor levels is unclear. Since progesterone levels do not fall prior to labor in women, oxytocin receptors in primates may be insensitive to down-regulation by progesterone. This possibility would best be addressed by antagonist studies in a non-human primate model, but to our knowledge these studies have not yet been performed.

X. OXYTOCIN–PROSTAGLANDIN INTERACTION

Numerous *in vivo* and *in vitro* studies suggest that an interaction between prostaglandins and oxytocin is necessary for the initiation and maintenance of spontaneous labor in women. Few studies have been performed with non-human primates. Human tissue studies have found that prostaglandin secretion, particularly $PGF_{2\alpha}$ by decidual cells is markedly stimulated by oxytocin (Fuchs et al., 1981). This and other data indicate that the decidua produces large amounts of prostaglandins in response to oxytocin stimulation. Husslein et al. (1981) have also presented convincing evidence supporting the necessity of an appropriate interaction for normal labor. These investigators found that induction of labor with oxytocin was only successful when the infusions were accompanied by a suitable response in uterine prostaglandin production. No similar studies have been performed with non-human primates.

Despite the likely role of oxytocin in stimulating prostaglandin secretion at the time of labor, some cases of preterm labor in women are not inhibited by oxytocin antagonist treatment (Akerlund et al., 1987). Furthermore, we have observed that uterine activity may increase following uterine surgery for catheter implantation in rhesus monkeys (personal observations). This type of uterine activity does not involve an increase in plasma oxytocin concentrations and is not inhibited by oxytocin antagonist treatment. In some of these animals prostaglandin levels may have become elevated under the influence of oxytocin stimulation before antagonist treatment was started. Alternatively, this type of uterine activity, and ultimately preterm labor, may be initiated in the absence of stimulation by oxytocin. Other factors, such as cytokine release in response to inflammation, may be of primary importance in the stimulation of prostaglandin production and onset of uterine activity after surgery. These observations further suggest that, although the onset of normal labor may involve oxytocin and prostaglandins acting together, preterm labor may be initiated by prostaglandins acting alone.

XI. CONCLUSIONS

The endocrine regulation of pregnancy has been investigated in a small number of non-human primate species with the majority of studies having been performed with either rhesus monkeys or baboons. Although there are differences between primate species, the salient features of the regulation of pregnancy are similar. This means that studies in these species are very useful for the investigation of human pregnancy and of the mechanisms involved in the initiation of labor. The ability to perform controlled experimental protocols with non-human primates has allowed observations found with human pregnancy to be tested. Such studies have determined that the primate fetus has only a minor influence in the control of the timing of labor onset. The use of chronically catheterized non-human primate preparations has resulted in the determination of the profile of prostaglandin concentrations in the amniotic fluid before and during the onset of labor. These studies have provided much supporting evidence for the view that the regulation of intrauterine prostaglandin production is of paramount importance in the initiation of labor in primates.

Evidence of a role for prostaglandins as the primary stimulatory factor in the initiation of labor is convincing in rhesus monkeys. As with human pregnancy, the local administration of prostaglandins leads to the induction of labor. Longitudinal studies in the rhesus monkey, however, have demonstrated that there is a steady rise in prostaglandin concentrations in the amniotic fluid before the onset of labor. These findings suggest that increasing concentrations of prostaglandins in the amniotic fluid ultimately lead to the initiation of labor in rhesus monkeys. Increasing intrauterine prostaglandin synthesis is most likely to be exclusively responsible for initiation of labor and delivery of the placenta following fetectomy or fetal death. Maternal factors may have the major role in the stimulation of prostaglandin synthesis. However, the nature of these factors remains to be determined.

Following the death of the primate fetus, the timing of labor and delivery shows a wide spread around normal term. This suggests that the fetus has a role in fine-tuning the timing of labor onset. Androgen secretion by the fetal adrenal glands is necessary for the maintenance of normal gestational estrogen concentrations. Increased production of fetal androgen leads to the rise in estrogen concentration that is observed prior to the onset of labor. Although this increase results in only a small rise in the ratio of estrogens to progesterone, this increase appears to be sufficient to stimulate an elevation in oxytocin receptor concentrations and gap junctions in uterine tissues.

The role of progesterone in the initiation of labor in primates is unclear. The experimental withdrawal of progesterone will initiate labor and delivery, however, uterine activity and cervical changes do not progress in the orderly manner observed at normal labor onset. Studies with rhesus monkeys indicate that uterine tissues may be sufficiently refractory to progesterone to allow the initiation of labor in the presence of gestational levels of progesterone.

Oxytocin concentrations rise at night during late gestation and reach peak levels during the early hours of darkness in rhesus monkeys. The magnitude of nocturnal peak oxytocin concentrations increases with advancing gestation. Elevated nocturnal oxytocin concentrations result in the generation of uterine activity episodes at night which also increase in size as term approaches. These episodes ultimately lead to onset of labor and delivery at night in rhesus monkeys. The development of nocturnal uterine activity episodes requires the presence of a suitable level of myometrial oxytocin receptors. This, in turn, is generated by the increasingly estrogenic environment produced by the fetus in late gestation. The rise in nocturnal oxytocin concentrations and the presence of elevated levels of oxytocin receptors allows the maintenance and enhancement of nocturnal uterine activity episodes near term. Once adequate levels of prostaglandins are attained within the intrauterine tissues, oxytocin-driven uterine activity progresses to labor and delivery in non-human primate species.

REFERENCES

Akerlund, M., Stromberg, P., Hauksson, A., Andersen, L.F., Lyndrup, J., Trojnar, J., & Melin, P. (1987). Inhibition of uterine contractions of premature labor with an oxytocin analogue. Results from a pilot study. Br. J. Obstet. Gynaecol. 94, 1040–1044.

Challis, J.R.G., Socol, M., & Murata, Y. (1980). Diurnal variations in maternal and fetal steroids in pregnant rhesus monkeys. Endocrinology 106, 1283–1288.

Chard, T., Hudson, C.N., Edwards, C.R.W., & Boyd, N.R.H. (1971). Release of oxytocin and vasopressin by the human fetus during labor. Nature 234, 352–354.

Chez, R.A., Hutchinson, D.L., Salazar, H., & Mintz, D.H. (1970). Some effects of fetal and maternal hopophysectomy in pregnancy. Am. J. Obstet. Gynecol. 108, 643–649.

Dawood, M.Y., Lauerson, N.H., & Trivedi, D. (1979). Studies on oxytocin in the baboon during pregnancy and delivery. Acta Endocrinol. 91, 704–718.

Ducsay, C.A., Cook, M.J., Walsh, S.W., & Novy, M.J. (1983). Circadian patterns and dexamethasone-induced changes in uterine activity in pregnant rhesus monkeys. Am. J. Obstet. Gynecol. 145, 389–396.

Fuchs, A.R., Husslein, P., & Fuchs, F. (1981). Oxytocin and the initiation of human labor. II. Stimulation of prostaglandin production in human decidua by oxytocin. Am. J. Obstet. Gynecol. 14, 694–697.

Fuchs, A.R., Fuchs, F., Husslein, P., Soloff, M.S., & Fernstrom, M.J. (1982). Oxytocin receptors and human parturition: A dual role for oxytocin in the initiation of labor. Science 215, 1396–1398.

Fuchs, A.R., Periyasamy, S., Alexandra, M., & Soloff, M.S. (1983). Correlation between oxytocin receptor concentration and responsiveness to oxytocin in the pregnant rat myometrium: Effects of ovarian steroids. Endocrinology 113, 742–749.

Fuchs, A.R., Romero, R., Keefe, D., Parra, M., & Oyarzun, E. (1991). Oxytocin secretion and human parturition: Pulse frequency and duration increases during spontaneous labor in women. Am. J. Obstet. Gynecol. 165, 1515–1523.

Garfield, R.E. (1988). Structural and functional studies of the control of myometrial contractility and labor. In: *The Onset of Labor: Cellular and Integrative Mechanisms* (McNellis, D., Challis, J., Macdonald, P., Nathanielsz, P., & Roberts, J., Eds.) pp. 55–80. Perinatology Press, Ithaca, NY.

Gravett, M.G., Witkin, S.S., & Novy, M.J. (1994). A nonhuman primate model for chorioamnionitis and preterm labor. Semin. Reprod. Endocrinol. 12, 246–262.

Haluska, G.J., & Novy, M.J. (1993). Hormonal modulation of uterine activity during primate parturition. Semin. Reprod. Endocrinol. 11, 261–271.

Haluska, G.J., Stanezyk, F.Z., Cook, M.J., & Novy, M.J. (1987). Temporal changes in uterine activity and prostaglandin response to RU486 in rhesus macaques in late gestation. Am. J. Obstet. Gynecol. 157, 1487–1495.

Haluska, G.J., West, N.B., Novy, M.J., & Brenner, R.M. (1990). Uterine estrogen receptors are increased by RU486 in late pregnant rhesus macaques but not after spontaneous labor. J. Clin. Endocrinol. Metab. 70, 181–186.

Hess, D.L., Spies, H.G., & Hendrickx, A.G. (1981). Diurnal steroid patterns during gestation in the rhesus macaque: Onset, daily variations and the effects of dexamethasone treatment. Biol. Reprod. 24, 609–616.

Hirst, J.J., West, N.B., Brenner, R.M., & Novy, M.J. (1990). Myometrial and decidual progesterone receptors during gestation, in labor and after RU486 treatment in rhesus monkeys. Serono Symposium, Uterine Contractility: Mechanisms of Control (Abstract #15).

Hirst, J.J., Haluska, G.J., Hess, D., Cook, M.J., & Novy, M.J. (1991). Comparison of plasma oxytocin and catecholamine concentrations with uterine activity in pregnant rhesus monkeys. J. Clin. Endocrinol. Metab. 73, 804–810.

Hirst, J.J., Chibbar, R., & Mitchell, B,F. (1993a). Role of oxytocin in the regulation of uterine activity during pregnancy and in the initiation of labor. Semin. Reprod. Endocrinol. 11, 219–233.

Hirst, J.J., Haluska, G.J., Cook, M.J., & Novy, M.J. (1993b). Plasma oxytocin and nocturnal uterine activity: Maternal but not fetal oxytocin concentrations increase during late pregnancy and delivery in rhesus monkeys. Am. J. Obstet. Gynecol. 169, 414–422.

Husslein, P., Fuchs, A.R., & Fuchs, F. (1981). Oxytocin and the initiation of human parturition. 1. Prostaglandin release during induction of labor with oxytocin. Am. J. Obstet. Gynecol. 141, 688–693.

Liggins, G.C., & Thorburn, G.D. (1994). Initiation of parturition. In: *Marshall's Physiology of Reproduction,* 4th ed., Vol. 3, *Pregnancy and Lactation,* Part Two (Lamming, G.E., Ed.) pp. 863–1002. Chapman and Hall, London.

Mitchell, B.F., Lye, S.J., Lukash, L., & Challis, J.R. (1986). Androstenedione metabolism in the late gestation sheep fetus. Endocrinology 118, 63–68.

Novy, M.J., & Haluska, G.J. (1988). Endocrine and paracrine control of parturition in rhesus monkeys. In: *The Onset of Labor: Cellular and Integrative Mechanisms* (McNellis, D., Challis, J., Macdonald, P., Nathanielsz, P., & Roberts, J., Eds.) pp. 321–334. Perinatology Press, Ithaca, NY.

Novy, M.J., & Walsh, S.W. (1983). Dexamethasone and estradiol treatment in pregnant macaques: Effects of gestational length, maternal plasma hormones, and fetal growth. Am. J. Obstet. Gynecol. 145, 920–926.

Novy, M.J., Cook, M.J., & Manaugh, L.C. (1974). Indomethacin blocks natural onset of parturition in primates. Am. J. Obstet. Gynecol. 118, 412–416.

Novy, M.J., Walsh, S.W., & Kittinger, G.W. (1977). Experimental fetal anencephaly in the rhesus monkey: Effect on gestational length and fetal and maternal plasma steroids. J. Clin. Endocrinol. Metab. 45, 1031–1038.

Pinto, R.M., Fisch, L., Schwarcz, R.L., & Montouri, E. (1964). Action of estradiol upon uterine contractility and the milk ejecting effect in pregnant women. Am. J. Obstet. Gynecol. 90, 99–107.

Resko, J.A., Ploem, J.G., & Stadelman, H.L. (1975). Estrogens in fetal and maternal plasma of the rhesus monkey. Endocrinology 97, 425–430.

Robinson, J.S., Chapman, R.L.K., Challis, J.R.G., Mitchell, M.D., & Thorburn, G.D. (1978). Administration of extra-amniotic arachidonic acid and the suppression of uterine prostaglandin synthesis during pregnancy in the rhesus monkey. J. Reprod. Fertil. 54, 369–373.

Romero, R., Soccia, B., Mazor, M., Wu, Y.K., & Benveniste, R. (1988). Evidence for a local change in the progesterone/estrogen ratio in human parturition at term. Am. J. Obstet. Gynecol. 159, 657–660.

Walsh, S.W. (1991). 5-hydroxyeicosatetraenoic acid, leukotriene C_4, and prostaglandin $F_{2\alpha}$ in amniotic fluid before and during term and preterm labor. Ann. NY Acad. Sci. 622, 341–354.

Walsh, S.W., Meyer, R.K., Wolf, R.C., & Friesen, H.G. (1977). Corpus luteum and fetoplacental functions in monkeys hypophysectomised during late pregnancy. Endocrinology 100, 845–850.

Walsh, S.W., Ducsay, C.A., & Novy, M.J. (1988). Circadian hormonal interactions among the mother, fetus and amniotic fluid. Am. J. Obstet. Gynecol. 150, 745–751.

Weiss, G., Steinetz, B.G., Deirshke, D.J., & Fritz, G. (1981). Relaxin secretion in the rhesus monkey. Biol. Reprod. 24, 565–567.

Wilson, L. Jr., Parsons, M.T., & Flouret, G. (1991). Forward shift in the initiation of the nocturnal estradiol surge in the pregnant baboon: Is this the genesis of labor? Am. J. Obstet. Gynecol. 165, 1487–1498.

THE PHYSIOLOGY OF HUMAN PARTURITION

Jane E. Mijovic and David M. Olson

Advances in Organ Biology
Volume 1, pages 89–119.

ISBN: 1-55938-639-8

ABSTRACT

Parturition is the process of giving birth. It results from a complex interplay of maternal and fetal factors due to the sequential maturation of an endocrine organ communication system. In sheep, there is conclusive evidence that this sequence begins at the level of the fetal brain where the regulatory mechanisms responsible for the changes in fetal adrenal glucocorticoid secretion are the initial signals for the onset of parturition. Increased fetal plasma cortisol levels result in a decrease in progesterone but an increase in estrogen in the maternal plasma and intrauterine tissues; this causes activation and stimulation of the myometrium, and effects labor. In primates, the fetal adrenal plays a different role—toward term it is responsible for providing increased estrogen precursors to the placenta; this changes the intrauterine estrogen:progesterone ratio in favor of estrogen, leading to the initiation of parturition. Estrogens stimulate the growth and interaction of the muscular proteins actin and myosin, an increase in the number, size, and permeability of gap junctions, and enable the generation of effective action potentials in the myometrium. Furthermore, under estrogen dominance there is an enhanced sensitivity of the myometrium to contractile agents and an increased synthesis and release of the uterotonins oxytocin and prostaglandins by the intrauterine tissues. Loss of inhibition is another mechanism which has been suggested as a mechanism for the initiation of contraction. This may take the form of the withdrawal of physiological regulators of the biosynthetic pathways of the uterotonins. Further to the development of rhythmical, sustained, and coordinated myometrial contractions, rupture of the fetal membranes, and ripening and dilatation of the cervix must occur for the successful delivery of the fetus. A glucocorticoid signal from the fetus may initiate a feed-forward cascade responsible for promoting all these events. Therefore, as in the sheep, the ultimate signals for birth may originate at the level of the primate fetal brain, initiating a multitude of fetal and maternal paracrine and endocrine processes resulting in the birth of a viable neonate.

I. INTRODUCTION: WHAT IS PARTURITION?

Parturition, the process of giving birth, is a highly orchestrated sequence of events resulting in the birth of the baby and the regeneration of normal (non-pregnant), cyclical uterine physiology. However, for such a simple definition, the process is very complicated and still eludes our complete understanding. As an example of its complexity, there are five separate physiological events which constitute parturition: fetal membrane rupture, cervical dilatation, myometrial contractility, placental separation, and uterine involution; each of which is an independent physiological action. In order for timely, normal birth to occur at term, all five physiological events must work in harmony. Because the vast majority of work is on one of these physiological events, the regulation of myometrial contractility, we will use it as the physiological endpoint in the majority of our discussions on the control of human parturition.

While the myometrium is never quiescent in terms of contractile activity, the transition from the contractile state of pregnancy to the contractile state of active labor is dramatic in both qualitative and quantitative terms. Long-duration, low amplitude contractions isolated to regionalized areas of the muscle characterize pregnancy, whereas the contractions of active labor are of short duration, high amplitude, and develop in a synchronous manner throughout the entire body of muscle. The transition of the uterine myometrium from the contractile state of pregnancy to the contractile state of labor, and the factors which lead up to this change, is the essence of the control of parturition, as seen in Figure 1. We will explore the control of human birth beginning with the concept that the fetus is the master of the timing of its own birth. The groundwork which has been laid through studies in animals, primarily sheep, will be explored briefly and used as a model for understanding the similarities and differences observed from primate and human studies in the translation of fetal neuro-endocrine maturation to control of maternal myometrial contractility.

ONSET OF LABOR

PREGNANCY

- Myometrium virtually quiescent
- Inexcitable
- Relatively unresonsive to uterotonic agonists
- Poorly coordinated

↓ **ACTIVATION**

LABOR

- High level of spontaneous activity
- Increased excitability
- Very responsive to uterotonic agonists
- Displays enhanced cell-to-cell coupling allowing generation of high frequency, intense, synchronous contractions of labor

Figure 1. The control of parturition is determined by the factors responsible for the transition of the uterine myometrium from the contractile state of pregnancy to the contractile state of labor.

II. THE FETUS AS MASTER OF ITS DESTINY

The concept that the fetus is aware of the proper time to initiate its own delivery dates from 460 B.C. The Greek philosopher Hippocrates suggested that the fetus could sense when the placenta was failing to keep up with its nutritional needs and when the time was opportune to exit the uterus. Two and one-half millennia later, as a result of experiments with sheep, Sir Joseph Barcroft concluded that the fetus monitors its own oxygen supply. When oxygen becomes deficient, mechanisms are activated to initiate parturition. Both ideas are similar in suggesting that fetal awareness of a deficiency of essential substances produces the need to be born (Swaab et al., 1976; Thorburn, 1994).

These views introduced the theory that the signal for birth originates at the level of the fetal brain. This was substantiated in later years by observations of disease conditions in pregnant animals. Kennedy and Holm's work from 1957 to 1964 established that prolonged pregnancy in cattle was associated with the failure of development of the fetal pituitary-adrenal axis. Kennedy reported a syndrome of prolonged gestation in Gurnsey and Jersey cattle. This syndrome was characterized by aplasia of the fetal anterior pituitary gland in which the adrenal medulla and cortex were distinguishable but small, and the cortical structure was undeveloped. Gestation in these animals ranged from 292–526 days, far exceeding the normal gestation length of 280 days. A second syndrome was observed in Holstein-Friesian cattle in which the fetuses failed to initiate labor and died *in utero* about 100 days past term. Holm concluded these calves suffered from a fetal addisonian syndrome (Kennedy, 1971; Holm, 1967).

A congenital cyclopian-type malformation associated with prolonged pregnancy was identified in fetuses of range sheep in Idaho in 1963. This teratogenic effect was a consequence of maternal ingestion of skunk cabbage (*Veratrum californicum*) on the fourteenth day of gestation. In these animals gestation lasted 200–250 days, again longer than the normal term of 150 days. The affected lambs had pituitary glands, but the neural connections with the hypothalamus were either missing or abnormal.

In 1965 Liggins developed a technique for ablating the pituitary gland of fetal sheep. When more than 70% of the pituitary gland was destroyed, pregnancy was prolonged. These fetuses had hypoplasia of the adrenal gland and retarded somatic development. Similar results were seen when the fetal hypothalamus was destroyed or the fetal pituitary stalk sectioned. Conversely, stimulating the fetal adrenal by infusion of adrenocorticotropin (ACTH) or direct administration of a glucocorticoid hormone into the fetal lamb *in utero* led to preterm delivery within a predictable number of days. The timing of parturition was unaltered by maternal hypophysectomy or by the administration of ACTH or synthetic glucocorticoids to the mother at similar doses and times in pregnancy.

In sheep, therefore, it is likely that the initial signal for parturition originates in the fetal hypothalamus. In the adult of most species, the hypothalamic paraventricular nucleus (PVN) secretes corticotrophin-releasing hormone (CRH) and arginine vasopressin (AVP) which regulates anterior pituitary corticotrope biosynthesis and ACTH release, which itself regulates adrenocortical glucocorticoid production. *In situ* hybridization methods demonstrated that messenger RNA (mRNA) for both CRH and AVP in the fetal PVN increased significantly between days 105 and 128 of gestation. Bilateral stereotaxically-placed lesions of the fetal ovine hypothalamic PVN prolonged gestation significantly and abolished the preparturient fetal plasma ACTH and cortisol rise. This method also showed that at day 130 the PVN signaled the onset of the expression (via ACTH) of fetal adrenal P_{450} steroid hydroxylase enzymes necessary for *de novo* cortisol synthesis. It is these signals for the activation of the fetal adrenal gland that can be considered the first fundamental step in the initiation of the endocrine cascade of ovine parturition (Challis et al., 1985).

A. Developing the Link Between Activation of the Fetal Brain and the Maternal Myometrium

In the sixteenth century the Italian anatomist Fabricius ab Aquapendente proposed that the chief agent of parturition was the muscular action of the uterus. The importance of the link between the maturation of the fetal hypothalamic-pituitary-adrenal axis and the initiation of labor can now be appreciated, at least in the sheep. Briefly, cortisol from the fetal adrenal brings about changes in maternal plasma hormone levels which lead to uterine activation and stimulation. Activation of the myometrium involves the development of an increased capability to contract (formation of gap junctions, an increase in receptors, changes in ion channels) and respond to the effectors (oxytocin, prostaglandins [PGs]) which stimulate the high frequency, high amplitude, and short duration contractions of labor. In the sheep, parturition is preceded by a large decrease in progesterone and a sharp rise in estrogens in the maternal blood. These steroids are synthesized by the placenta throughout the majority of gestation. Mean fetal plasma cortisol levels are highest during the final 2–3 days of intrauterine life. At these levels cortisol induces placental 17α-hydroxylase synthesis which converts progesterone to 17α-hydroxyprogesterone, an estrogen substrate. Therefore, maternal plasma progesterone levels decrease 1–2 days before labor onset and estrogen levels sharply increase in the 36 hours leading up to labor. This promotes uterine activation, increased synthesis of uterotonic contractile agents (uterotonins), such as $PGF_{2\alpha,}$ and results in myometrial contractions. These changes are induced prematurely in those ewes where *in vivo* administration of ACTH or synthetic glucocorticoid (dexamethasone) to the fetus results in premature parturition.

B. The Human and Non-human Primate Evidence for Fetal Control of Birth

Evidence for the role of the fetus in the initiation of parturition in the human is more circumspect. In fact, for obvious ethical reasons, much of our knowledge concerning the physiology of human parturition is often from purely anecdotal sources. Because of this, it is often pertinent to refer to studies using animal models. While this is recognized as a necessity, its physiological significance is not always clear; therefore, in these discussions it will be kept to a minimum.

Observations on the Fetal Control of Birth in the Human

In 1933 an English obstetrician, Malpas, reported a series of cases of prolonged pregnancy in humans associated with fetal anencephaly. He observed that the front portions of the fetal brain failed to develop and concluded the timing of the onset of labor is determined by the fetus. He suggested that the fetal adrenal, pituitary, and nervous system, in combination, trigger the neuro-muscular mechanisms effecting labor (Malpas, 1933).

Further observations on the effects of anencephaly on gestational length in humans have been carried out. Where a large number of patients have been monitored, the mean length of gestation of the anencephalic group was not found to be significantly different from that of the control population. However, in the anencephalic group there was a much wider variation around the mean, the second and third stages of labor were significantly longer, and manual extraction of the placenta was more frequent. These latter two observations may be due to an influence of fetal neurohypophyseal peptides on myometrial contractility and the duration of labor that is absent in the anencephalic group. These studies suggest the human fetus plays a role different from its ovine counterpart in determining the timing of labor onset and its maintenance thereafter. These and other studies over the past 25 years have led to the firm conclusion that the fetal brain plays a critical role in triggering the onset of labor (for a review see Challis and Olson, 1988).

However, in sharp contrast to the fetal sheep, *in vivo* research does not suggest that primate fetal adrenal cortisol has an analogous role in the timing of parturition. In both monkeys and humans, exogenous glucocorticoids fail to induce parturition, except in some patients already classified as post-term. The amounts of glucocorticoids given intra-amniotically to such patients are large and their effect is inconsistent. In addition, dexamethasone administration results in a decrease of maternal estrogen levels. In humans as in the sheep, increasing maternal estrogen concentrations are implicated in uterine activation and uterotonin synthesis. These data suggest fetal cortisol is unlikely to serve as the physiological trigger to parturition in humans (Casey et al., 1985). However, closer analysis of more recent research may yet reveal an essential role for fetal glucocorticoids in these processes.

The Human Fetal Adrenal Gland

It is established that the human fetal adrenal is involved in parturition through the synthesis and secretion of steroids such as the C_{19} estrogen precursors, dehydroepiandrosterone (DHEA), and its sulfated form, DHEAS. The placenta is the major site of steroid production after the sixth to ninth week of human pregnancy. The primate placenta cannot synthesize estrogens *de novo* from acetate as it lacks the 17α-hydroxylase/C17-20 lyase enzymatic system responsible for this. However, it is capable of converting C_{19} steroids into estrogen. In human and nonhuman primates maternal plasma concentrations of unconjugated estrogens rise progressively with time of gestation reaching peak levels at term. Evidence suggesting this estrogen production is closely related to the secretion of DHEAS from the fetal adrenal gland has been reviewed several times (Pepe and Albrecht, 1990a, 1990b).

The fetal adrenal gland of the human, like that of the fetal rhesus monkey, shows functional specialization. There is an outer adult-type definitive zone or neocortex and an inner fetal cortical zone which comprises between 80 and 90% of the gland during the majority of gestation. The fetal zone exhibits a period of rapid growth from day 150 of gestation. It secretes primarily the C_{19} steroid precursors of estrogen in increasing quantities throughout gestation in accordance with its increasing mass. The adult cortex secretes primarily cortisol; maturation of the adult cortex occurs very late in gestation and results in an increase in fetal plasma cortisol levels. At mid to late gestation fetal adrenal size and weight in anencephalic human fetuses are considerably reduced when compared with normal. This suggests factors of fetal pituitary origin are important to fetal adrenal growth at this time. These include ACTH and other proopiomelanocortin-derived peptides (POMC) as well as fetal and placental growth factors. Anterior pituitary POMC expression is regulated by the synthesis of neuropeptide releasing factors such as CRH and AVP.

In human fetuses CRH- and AVP-like immunoreactivities were found to be present in hypothalamic tissue by 12 weeks, and the concentration of bioactive CRH rose between 12 and 27 weeks. There is also evidence for the production of placental CRH and POMC-derived peptides (Riley and Challis, 1991). Studies show amniotic fluid and maternal and fetal plasma immunoreactive CRH (IR-CRH) rise steadily from mid-second trimester to 35 weeks, after which time they increase rapidly to term (ca. 40 weeks). These increasing concentrations correlate with increasing

IR-CRH levels in placental extracts. However, there are no further changes with labor onset. Glucocorticoids increase placental CRH gene expression and mRNA levels; CRH enhances PG production by fetal membranes, the myometrium, and the placenta and exerts a priming and potentiating effect on the myometrial contractile response to the uterotonin, oxytocin. Further, the receptor affinity for CRH in intrauterine tissues increases with gestational age. Based on these observations it is possible that late in gestation when the human fetal adrenal adult cortex has matured and begins to secrete cortisol, the increased glucocorticoid concentrations could then act to stimulate placental CRH expression. This would initiate an intraplacental positive cascade that stimulates prostaglandin production by paracrine/autocrine interactions and modulates the sensitivity of the myometrium to contractile effectors. Once concentrations of CRH in amniotic fluid and maternal and fetal plasma increase, they may further drive the pituitary-adrenal axis in the fetus and possibly the mother. When this process has been initiated, feedback control must be diminished to allow propagation of this set of signals to stimulate the onset and maintenance of labor. Alternatively, oxytocin and PGs maintain the intraplacental cascade; oxytocin stimulates human placental CRH release, and PGs stimulate ACTH and cortisol release from the pituitary-adrenal axis of fetal sheep. In this case the timing of the maturation of the fetal adrenal adult definitive zone plays a critical role in signaling the onset of parturition in primates. The failure of exogenous glucocorticoid administration to induce premature labor in primates might be due to the fact that treatments were given too early in gestation when negative feedback control was still in operation. This together with the resultant decrease in estrogen precursors would simply serve to prolong gestation.

Primate Fetal Adrenal-pituitary Maturation

In addition, the primate fetus has a role in the timing of the maturation of its own hypotholamic-pituitary-adrenal axis. During the majority of intrauterine life the primate fetus has limited capacity to produce cortisol. The fetus obtains this hormone by transplacental transfer from the mother. At mid-gestation 100% of fetal serum cortisol is derived from the mother while at term less than 50% originates from hormone produced by the maternal adrenal gland. The enzyme, 11β-hydroxysteroid dehydrogenase, catalyzes the oxidation and reduction of cortisol and cortisone, respectively, and is present in the placenta and fetal membranes of humans throughout gestation. Placental cortisol metabolism changes with advancing gestation from reduction at midgestation to oxidation near term. The increase in placental cortisol oxidation is regulated by estrogen, the production of which increases to term. Cortisone has no biological activity unless it is converted to cortisol.

Pepe and Albrecht (1990a, 1990b) have proposed the following model for primate fetal pituitary-adrenal maturation. At midgestation cortisol is the principal corticosteroid arriving within the fetal circulation; fetal pituitary activity and thus ACTH release is suppressed, limiting the growth of the adult definitive zone of the fetal adrenal including development of the enzyme systems for producing glucocorticoids. The growth of the fetal cortex and consequent DHEA and DHEAS production continues under the influence of fetal and placental growth factors. With advancing gestation, under the influence of increasing estrogen dictated by increasing fetal adrenal C_{19} precursors, transplacental metabolism is changed, and the oxidation of cortisol to cortisone is increased. Concentrations of maternal cortisol in the fetal circulation drop, resulting in increased production and release of ACTH of fetal pituitary origin, maturation of the fetal adrenal adult definitive zone and *de novo* cortisol production.

In this way the primate fetus, initially through secretion of estrogen precursors by the adrenal gland and then ACTH and cortisol from the pituitary-adrenal axis, has the potential to exert greater control over the timing of its own birth than was previously thought. It is noteworthy that cortisol also acts to induce systems to bring about maturation of many fetal organ systems, especially the lung. Thus cortisol acts as a means of coordinating fetal organ maturation with the timing of birth. Furthermore, fetal lung surfactant in the amniotic fluid can stimulate PG output from the amnion and is therefore another way of linking fetal lung maturity to a trigger for labor onset. While much of this scheme is still speculative and active investigation is continuing in this field, these data, nevertheless, support the contention that activation of the fetal hypothalamic-pituitary-adrenal axis is responsible for the initiation of parturition in primates as well as in sheep, as seen in Figure 2.

C. Evidence that the Fetal Endocrine Message Regulates the Timing of Birth

In the rhesus monkey spontaneous delivery is preceded by rising concentrations of DHEAS in fetal blood, but not in maternal blood or

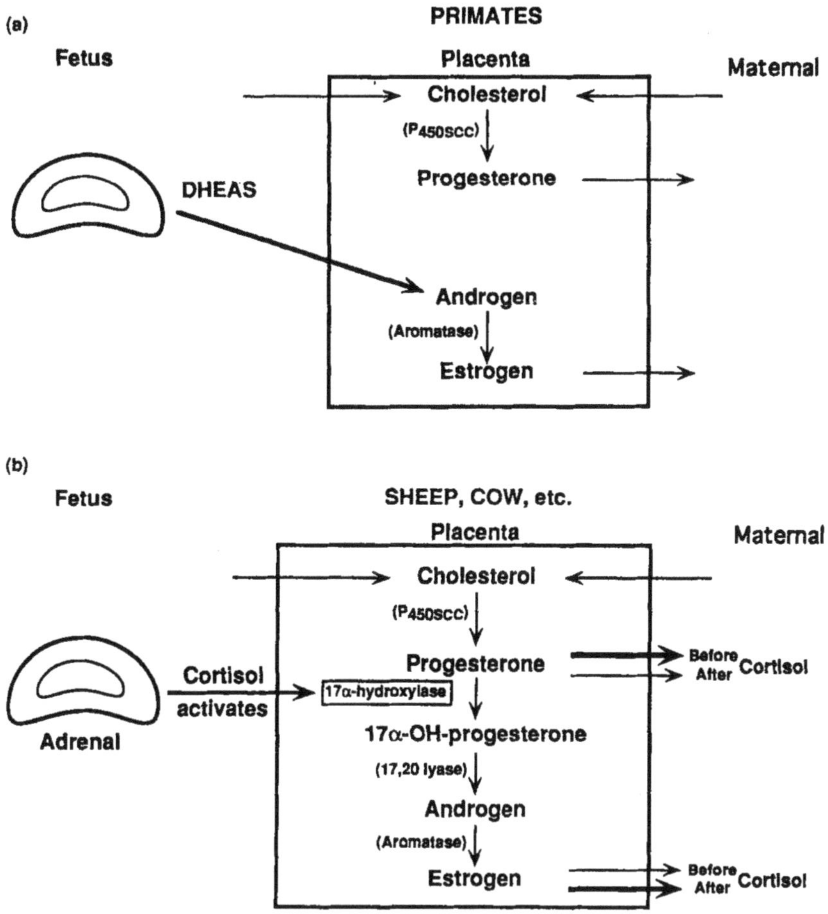

Figure 2. Activation of the fetal hypothalmic-pituitary-adrenal axis is responsible for the initiation of parturition in primates as well as sheep. In primates fetal adrenal DHEAS may play a major role in this process, while in sheep fetal control is implied as the trigger for labor.

amniotic fluid. Furthermore, *in vivo* data support the concept of rapid fetal adrenal maturation before birth, and fetectomy in rhesus monkeys and baboons, fetal decapitation, fetal death, or suppression of the fetal adrenocortical axis with glucocorticoids in rhesus monkeys causes peripheral maternal estrogen levels to fall and a significant increase in gestational length. Conversely, an increase in the production of fetal adrenal androgens and consequently maternal estrogens similar to that seen at spontaneous term labor (184 days) has also been found in pregnant baboons at 155–165 days of gestation treated to cause fetal hypoxic stress which would normally result in preterm birth at this time. Furthermore, circulating maternal estrogens rise after administration of ACTH or DHEAS to the fetus, and administration of androstenedione to the rhesus monkey fetus causes a switch in myometrial activity from contractures to contractions and an increase in circulating maternal estrogen levels. This further suggests that the fetoplacental unit plays an important part in the signaling process in preparation for parturition. In humans it is also likely that steroidal precursors for estrogen production arise in the fetus as pregnancy with an anencephalic fetus is associated with subnormal levels of urinary estrogen, estriol secretion, and fetal adrenal hypoplasia; and low estriol secretion is also observed after fetal death and after suppression of the fetal adrenocortical axis with synthetic glucocorticoid. While DHEAS and androstenedione administered to the fetus cause increases in uterine contractile activity, these results cannot be replicated with estrogen administration in the non-human primate. This may be due to serum binding or metabolism of the estrogen, or it may reflect a local, placental effect of estrogen formed from the androgen which cannot be replicated by exogenous estrogen administration. Alternatively, androgen may have its own intrinsic effect upon uterine contraction. More investigation must be performed to resolve this discrepancy.

III. THE ROLES OF MATERNAL SEX STEROIDS IN THE MAINTENANCE AND TERMINATION OF PREGNANCY

Many studies have defined a reduction in the concentrations of circulating maternal progesterone as the point at which the initiation of parturition occurs. In 1956 Csapo suggested that delivery could not take place before a fall in maternal progesterone levels because progesterone blocked the activity of the myometrium. In species such as the rabbit, the corpora lutea are maintained throughout pregnancy and provide progesterone to maintain pregnancy. Ablation of the corpora lutea or ovaries leads to abortion. Similarly, normal parturition in rabbits can be prevented by exogenous progesterone administration.

A. The Progesterone Block Theory

In 1976 Csapo expanded the progesterone-block theory to the seesaw theory of pregnancy maintenance and termination. He proposed that the myometrial regulators, progesterone and PGs, were opposing forces in the control of the two ends of the seesaw. It was suggested that under progesterone dominance, the myometrium bound calcium to the cell membrane and the sarcoplasmic recticulum. The contractile force of parturition is generated by calcium-activated interaction between actomyosin and adenosine triphosphate (ATP). This maintained uterine quiescence during pregnancy. However, under the increasing production and effectiveness of PG at term and declining progesterone dominance, the balance of the seesaw was tipped so that $PGF_{2\alpha}$ promoted the influx of activator calcium from and through the membrane. The theory suggested that PGs could not promote uterine contractions when progesterone was so strong, it controlled myometrial function. Csapo also included a role for estrogen in his theory, suggesting that estrogen domination of the hormonal environment promoted intrauterine changes which favored myometrial contractility (Csapo, 1976).

The non-pregnant sheep uterus conforms to the classic progesterone block theory of Csapo. Progesterone inhibits the sheep myometrium and blocks the action of oxytocin and PGs. Furthermore, in the pregnant sheep there is a decrease in peripheral progesterone 1–2 days before labor onset and a switch from progesterone to estrogen domination. These changes occur in response to fetal signals and thus are parts of the endocrine cascade leading to labor.

B. The Estrogen: Progesterone Ratio in Late Pregnancy

There is essentially no evidence that systemic maternal progesterone withdrawal is a prerequisite for the initiation of primate parturition. In humans, maternal peripheral progesterone levels increase progressively toward term. However, as the antiprogestin, RU486, a receptor blocker, will induce abortion in women, a role for progesterone acting via its receptor and resulting in uterine quiescence is envisioned; hence, a change in progesterone to estrogen effect is purportedly associated with primate parturition. So, how is that effect manifest? Estrogen concentrations also increase progressively prior to labor in women, but, with equally increasing progesterone levels, there is no change in the systemic estrogen:progesterone ratio in favor of estrogen. Therefore, investigators have looked for more subtle or local changes in the estrogen:progesterone ratio in primates which may yield the same systemic effects observed in animals. One such subtlety may reflect the time of day at which steroids are determined. Both steroids are known to be secreted in a circadian rhythm at all stages of gestation in primates. During the last

10–12 days of pregnancy there is a forward shift in the initiation of the nocturnal estradiol surge so that it occurs before the progesterone surge. This can result in a temporary (3–5 h), but effective, increase in the estrogen:progesterone ratio. This forward shift coincides with the beginning of nocturnal uterine activity and may be a signal for the initiation of parturition.

It is apparent that estrogen formed within the placenta is involved in the regulation of progesterone formed by this tissue during primate pregnancy. Administration of the antiestrogen, MER-25, a receptor antagonist, to baboons in the last one-third of gestation was associated with a 50% decline in serum concentration and production rate of progesterone. These data suggest estrogen regulates the biosynthesis of placental progesterone.

Pepe and Albrecht have shown that a regulatory system exists *in utero* at mid-gestation in which there is negative feedback control of placental estrogen on the secretion of the fetal adrenal DHEAS, possibly by attenuating responsivity to trophic peptides (ACTH). During late gestation this effect is lost resulting in the increased fetal DHEAS and maternal estrogens seen at term. Presumably this does not result in a parallel increase in placental progesterone, as it is proposed that normally the placenta is maximally stimulated by estrogen. Therefore, these processes would lead to an increase in the estrogen:progesterone ratio at term. As placental estrogen formation is dependent on fetal androgen precursors from the adrenal gland, this process is ultimately directed by the fetus.

Additionally, the concept has developed that local intrauterine tissue changes in steroid concentrations occur prior to or during parturition which are not reflected by changes in peripheral plasma concentrations. Several studies in this regard have been carried out by Mitchell's laboratory. The activities of several enzymes change during the onset of labor in decidua and chorion which suggest the possibility that relatively more estrogen and less progesterone may be present in these crucial tissues. The activity of 3β-hydroxysteroid dehydrogenase, which converts pregnenolone to progesterone in decidua decreases during spontaneous labor. Estrone sulfotransferase, which catalyzes the release of unconjugated estrone from estrone sulfate, displays an increase in its activity in chorion and decidua during labor. Further, the reductive activity of the enzyme, 17β, 20α-hydroxysteroid dehydrogenase, which "activates" estrone to estradiol and "inactivates" progesterone to 20α-dihydroprogesterone in amnion and chorion, is increased in labor. The net result of these enzyme activity changes suggests a local decrease in the progesterone:estrogen ratios and a local withdrawal of progesterone at birth. However, direct determinations of estrone, estradiol, and progesterone in amnion, chorion, and decidua showed that no change in the absolute concentrations of steroids or in the estrogen:progesterone ratio occurred with labor onset. Hence while the potential has been demonstrated, no evidence exists for an actual systemic or local withdrawal of progesterone at term or with labor (for a review see Hirst et al., 1993).

C. The Role of Steroid Hormone Receptors

A change in the estrogen:progesterone ratio may also occur at the receptor level. Classically, estrogen and progesterone exert their effects by binding first to a cyotoplasmic macromolecule to form a receptor steroid complex. This complex undergoes translocation to specific nuclear sites where it stimulates RNA synthesis which is responsible for modifying cell growth and function. Thus, it is the nuclear receptors that are biologically active. The cytoplasmic receptors are replenished by recycling of the receptor macromolecule and by *de novo* synthesis.

In rat intrauterine tissue, cytosol, and nuclear estrogen receptor concentrations increase abruptly and are maximal at labor. Cytosol progesterone receptor concentrations increase throughout pregnancy until term, whereas nuclear progesterone receptor concentrations fall prior to delivery, paralleling the changes in steroid concentrations at this time. In human myometrium, low or absent levels of progesterone and estrogen receptor have been shown throughout pregnancy. These low concentrations may be due to antagonism of receptor synthesis by high progesterone concentrations, however, the estrogen:progesterone ratio may also be important in this regard. In the lower myometrial segment of women at spontaneous labor or undergoing elective cesarean section without labor, estrogen and progesterone receptors appeared equally distributed between the cytosol and the nucleus. No differences in estrogen receptors were evident between the two groups, but a decreased progesterone receptor concentration was seen in the labored tissues. Estrogen and progesterone receptors also exist in the fetal membranes and decidua. Estrogen receptor mRNA concentrations increase three to fourfold around the time of labor onset, whereas progesterone receptor mRNA remains unchanged. These data suggest that an altered action of estrogen relative to progesterone may be observed with the onset of labor in humans as a consequence of increased intrauterine tissue estrogen receptor relative to progesterone receptor (How et al., 1994).

It is now becoming increasingly apparent that sex steroids may alter cell function in other ways than through classic receptor mechanisms. These include binding to the cell membrane thereby affecting its structure and fluidity and altering the interactions of many agonists-membrane receptor-second messenger systems. In this context a relaxation action of progestins on rat myometrium via a membrane effect which inhibits extracellular calcium influx by calcium channels is known to exist. Progesterone may act by modulating calcium channel opening. Therefore, extrapolating data about steroid activity from receptor studies must be viewed with care considering the emergence of evidence suggesting non-receptor mediated effects.

IV. EFFECTS OF STEROIDS ON UTERINE MUSCLE ACTIVATION AND STIMULATION

A. Myometrial Activation

Estrogen and progesterone have no direct effect on contractility, but regulate it through their actions on protein synthesis, on synthesis of intracellular and cell surface receptors, on phospholipid membrane structure and uterotonin synthesis. Progesterone exerts its blocking action on excitation and conduction mechanisms. The ability of the myometrium to conduct action potentials is inhibited, thus activity is not propagated but is confined to local areas only. In the absence of coordinated propagated activity there can be no significant development of intrauterine pressure. Further, progesterone domination of the uterus serves to maintain smooth muscle intracellular calcium at a low level, and this prevents a response to stimulators of myometrial contractility.

It is well established that a change in the steroid environment from progesterone to estrogen dominance causes activation and stimulation of the myometrium and effects labor. The essential phenomenon of uterine contraction is the interaction between the two muscular proteins, myosin and actin. Myosin is composed of heavy and light chains which interact with actin causing ATP hydrolysis and the generation of force. The interaction between these two proteins is promoted by the enzymatic phosphorylation of the myosin light chain, catalyzed by the enzyme myosin light chain kinase, which is activated by a calcium-calmodulin complex. Estrogens stimulate growth and actomyosin formation in the human uterus and are essential for the maintenance of an effective membrane potential to enable generation of action potentials (Word, 1995).

In order for smooth muscle contraction to spread from cell to cell and produce a synchronized uterine contraction, cell-to-cell communication through gap junctions is necessary. The number and size of gap junction contacts in humans increases at the surface of myometrial cells immediately before the onset of labor. They are present during preterm labor and absent when delivery is prevented. Furthermore, studies also show that if the junctions are prevented from forming by manipulating hormones, labor and delivery are either delayed or prevented. The use of estrogen and progesterone agonists and antagonists shows that the formation of myometrial gap junctions occurs as a result of the action of estrogenic hormones while progesterone opposes this stimulatory action (Burghardt et al., 1993). The most abundant myometrial gap junction protein has been identified as connexin-43 (Cx-43). In both pregnant and non-pregnant rats the steroid environment is important for modulating the steady state transcripts encoding Cx-43; estrogen acts to increase the level of transcripts, and progesterone can both block and reverse this action of estrogen. Further, estrogens may affect gap junction permeability as isolated myometrial cells primed with 17β-estradiol compared to untreated controls showed increased intracellular communication. In sheep, the timing of myometrial gap junction formation and function can be correlated with an increase in the ratio of circulating estrogen to progesterone. In the human myometrium, a local progesterone withdrawal is implicated in this process (for a review see Lye, 1994).

B. Myometrial Uterotonins

The increased sensitivity of the myometrium to contractile agents is determined by increased receptor numbers and enhanced receptor coupling to second-messenger systems. In sheep the change in the steroid environment to estrogen dominance promotes the formation of receptors for oxytocin and PGs. In women estrogen treatment in late pregnancy increases oxytocin sensitivity by stimulating an increase in oxytocin receptor concentrations. PGs can be identified as positive feedback agents in this cascade as they further enhance estrogen-induced expression of oxytocin receptors (for reviews see Fuchs, 1995; Olson et al., 1995).

In addition, estrogens stimulate endocrine and paracrine production of the oxytocin peptide. The release and synthesis of oxytocin in the hypothalamus is increased due to stimulation of oxytocin gene expression by estrogen. Furthermore, in human amnion, chorion and decidua, estrogen stimulated a fourfold increase in oxytocin mRNA *in vitro* which is an increase similar to that seen in tissues obtained around the time of spontaneous parturition relative to those obtained earlier in gestation. Progesterone had little effect on this response alone and inhibited the estrogen response in some experiments. Oxytocin stimulates phosphatidylinositiol hydrolysis in myometrial cells resulting in increased inositol triphosphate generation which is a second messenger step leading to myometrial contraction. Diacylglycerol is also a product of this hydrolysis; the action of cellular lipases on diacylglycerol releases arachidonic acid which is the substrate of PGs. Also, the release of intracellular calcium initiated by phosphatidylinositol hydrolysis can cause further mobilization of arachidonic acid from phospholipid stores as the phopholipases responsible for this reaction are Ca^{2+} dependent. Furthermore, oxytocin has been shown to induce the production of another uterotonin, endothelin-I, in decidua. Comparable to all these stimulatory agents, endothelin-I induces calcium influx into myometrial cells and is equipotent to oxytocin in stimulating uterine contractions.

There is other *in vitro* evidence that the production of PGs by human intrauterine tissues is regulated hormonally. In general, estrogen stimulates PG production while progesterone can either antagonize or augment estrogen action. In culture, estrogen has been shown to stimulate the output of PGE_2 from decidual cell preparations obtained from patients at elective cesarean section. In the myometrium, dexamethasone decreased prostacyclin synthesis by 90%. Prostacyclin (PGI_2) causes myometrial relaxation, and the increase in fetal glucocorticoid production at term described previously may have a further role in decreasing PGI_2 synthesis and causing a switch to the production of stimulatory $PGF_{2\alpha}$ and PGE_2.

Estradiol and progesterone have no effect on the expression of the two isoforms of prostaglandin endoperoxide H synthase (PGHS), the enzyme catalyzing the rate limiting step of the conversion of arachidonic acid to PGE_2 in cultured amnion cells. A rate-limiting step in metabolism of PGs is the 15-prostaglandin dehydrogenase enzyme (PGDH) which is found in most intrauterine tissues, especially chorion. The activity of this enzyme is stimulated by progesterone. Therefore, under the progesterone domination of pregnancy, PGDH enzyme levels will be high, reducing the bioactivity of any PGs formed. Local withdrawal of progesterone at term may reduce PDGH activity, thus increasing PG bioactivity. In support of this are the observations of women treated early in pregnancy with the progesterone receptor antagonist, RU486. Such treatment is associated with a dramatic decrease in decidual PGDH activity and increased spontaneous uterine contractility (Hansen, 1976).

V. SYNTHESIS AND ACTION OF MYOMETRIAL CONTRACTILE STIMULATORS

A. Oxytocin

Perhaps the most appealing evidence which reveals the dual physiological roles of uterine activation and stimulation is that of increasing myometrial responsiveness to oxytocin in late gestation. Hirst and his colleagues studied the maternal plasma concentrations of oxytocin during late gestation in rhesus monkeys. They observed an increase in the amount released from day 130 to day 156 gestation (term = day 167), and that the release of oxytocin occurred during the dark hours, between 2100 h and 0300 h. The importance of this finding is related to the observation that rhesus monkey myometrial contractility displays a nocturnal increase in contractile activity during late gestation when contractures become contractions. With succeeding nights, the intensity and duration of the contractions increase, in parallel with increased oxytocin levels in plasma. To demonstrate the correlation between nocturnal uterine activity and nocturnal oxytocin release, Atosiban, an oxytocin receptor antagonist was administered. It blocked the nocturnal increase in uterine contractile activity (contractions), but upon cessation of administration, nocturnal uterine activity returned. Daytime uterine activity and oxytocin levels were baseline throughout this period of late gestation.

To further test whether uterine sensitivity to oxytocin varied, oxytocin was infused at different times of the day into pregnant rhesus monkeys. The researchers observed that the myometrial contractile responsiveness was greatest at night when compared to afternoon or morning. It is at night when the fetal DHEAS levels are highest, and DHEAS administration to the fetus causes a switch from contractures to contractions. Hence, it is probable that raising fetal DHEAS levels directly or indirectly through estrogen formation activates the uterus, while enhanced oxytocin release, perhaps again stimulated through elevated estrogen production, causes the uterus to develop contractions (see Olson et al., 1995 for review).

Recent work has demonstrated the potential of intrauterine tissues to produce oxytocin and their estrogen sensitivity in this regard. It has been shown that the rat uterus had large increases in oxytocin mRNA abundance on days 18 and 21 (birth = day 22). The intrauterine sources of oxytocin are endometrium, placenta, and amnion, with the endometrium being the probable site of late pregnancy changes in gene expression. When treated with steroids, progesterone alone had no effect upon rat uterine oxytocin mRNA expression and estrogen raised the expression slightly, but significantly. However, when both estrogen and progesterone were administered together, a synergistic increase in oxytocin mRNA occurred. This increase was not as large as the increase observed on day 21 of pregnancy, however, suggesting other potential regulators of rat uterine oxytocin mRNA expression (Alexandrova and Soloff, 1980). Given the small number of hypothalamic oxytocin neurons which contain estrogen receptors, it is possible that uterine oxytocin is the primary site of estrogen-mediated increases in oxytocin production. Chibbar found that oxytocin mRNA is expressed in human intrauterine tissues, with most coming from decidua and lesser amounts in chorion and amnion, while the placenta does not have detectable levels. The expression in decidua increases with the onset of labor at term. Incubation of choriodecidua tissue with estrogen promoted a threefold increase in oxytocin mRNA expression while progesterone had no effect.

These intrauterine interactions between estrogen and oxytocin expression suggest the possibility of a paracrine control of oxytocin and PG production in the uterine tissue. Activation of the primate fetal adrenal may lead to increased production of fetal DHEAS which is aromatized to estrogen in the placenta. This estrogen may lead to enhanced decidual expression, production, and storage of oxytocin. At the same time, increasing estrogen may enhance production of $PGF_{2\alpha}$ from the decidua. This $PGF_{2\alpha}$ may stimulate the release of oxytocin, which in other species is known to trigger the production of more $PGF_{2\alpha}$. Hence, two uterine contractile stimulants may have a positive feed-forward relationship whereby they stimulate the production of one another in addition to stimulation of the myometrium. This possibility needs to be tested.

Another question which has been addressed, but not completely answered, is whether oxytocin is actually the uterine stimulant which leads to labor initiation. The use of oxytocin receptor antagonists has shown that their administration prevents the nocturnal switch from contractures to contractions in rhesus monkeys and baboons. In guinea pigs the administration of an oxytocin receptor antagonist decreased uterine electromyographic (EMG) activity and prolonged the expulsive (second) phase of labor. This resulted in a delay of fetal delivery and increased fetal mortality. However, maximum uterine activity occurred at the expected time and the transition of EMG activity from a prepartum to a postpartum pattern was normal. These data suggest that oxytocin is not involved with the mechanism of the onset and timing of labor, but that it does contribute to the expulsive phase of labor. Similar studies must be performed in primates, and the potential interaction between PG and oxytocin synthesis needs to be included.

B. Prostaglandins

The other major uterine stimulant is PG. The sites of PG synthesis in the human and higher primates are the intrauterine tissues, amnion producing primarily PGE_2 and decidua producing $PGF_{2\alpha}$ plus other PGs (for reviews see Olson et al., 1993). The lines of evidence supporting the theory that PGs are important in the initiation and maintenance of labor include the administration of PG from mid-pregnancy to term causes labor; that blocking PG synthesis delays labor onset, reduces contractions, and prolongs the process of labor; that PG levels in maternal plasma, urine; and amniotic fluid rise with labor progression; and that PGs stimulate uterine contractility *in vitro*.

While evidence may be strongly supportive of a role of PGs in labor initiation in sheep and even rhesus monkeys, the data are not as clearcut in humans. Indeed, it has been suggested that PGs may actually be a consequence and not a cause of labor in humans. Principal among the difficulties in humans is the fact that the major increase in amniotic fluid and maternal plasma PGs occurs after labor has started rather than before. Through the gradually opening cervix, the fetal membranes and attached decidua may come into contact with vaginal fluids and the bacteria contained therein. This, combined with decidual necrosis, leads to enhanced local production of PGs which is reflected by a gradient of PG concentrations within the amniotic fluid. Consequently, the observed large increase in PG concentrations in amniotic fluid may occur as a function of labor.

Although this evidence explains much of the large increase observed in amniotic fluid and, perhaps via absorption by the vagina, maternal plasma increases in PGs during second stage labor, it does not explain the pre-labor increases in amniotic fluid PGs which several laboratories have observed. These studies, using transabdominal or transcervical amniocentesis, sampled patients before the onset of labor as established by several clinical criteria. Even though transcervical amniocentesis can be criticized because amniotic fluid can come into contact with vaginal secretions, it does not account for the changes observed in late gestation amniotic fluid PG concentrations as the procedure was performed identically in each patient. Preterm increases in sheep and rhesus monkey amniotic fluid PGs are well documented and have been correlated with increasing myometrial contractility. The questions which must be addressed in future studies are what is the concentration of PG at the myometrium, and what is the critical concentration needed to stimulate uterine contractility?

The direction of a great deal of work now is to explore the mechanisms responsible for changes in PG production in late gestation. There does not appear to be any circadian rhythm to the pattern of PG production, therefore it is unlikely that acute changes in DHEAS or estrogen play a prominent role in PG regulation.

The two steps of the pathway leading to PGs which are most amenable to control are the deacylation, or release, of arachidonic acid and its conversion to endoperoxide. Two specific areas in relation to this process are receiving a great deal of attention recently. The first is the identification of an extracellular (Type II) phospholipase A_2 (PLA_2-II) which is the most abundant PLA_2 present in human gestational tissues as shown by Rice and his laboratory group. Importantly, its immunoreactive and catalytic activity increase in amnion, choriodecidua, and placenta during labor, but not after labor. There is no corresponding change in mRNA, suggesting an acute regulation at a post-transcriptional site. None of the other descriptions of phospholipase have demonstrated a change in catalytic activity at term, hence the justification of excitement over these findings.

The second area of attention is the control of PG synthesis at the prostaglandin endoperoxide H synthase (PGHS) or cyclooxygenase step. Recently it has been shown that human amnion PGHS enzyme activity increases in late gestation, before the onset of labor, which parallels the changes in amniotic fluid PGs seen in other studies. Two forms of PGHS exist, a constitutive form PGHS-1, and an inducible form PGHS-2. PGHS-1 and -2 mRNA and immunoreactive proteins have been identified in amnion and decidua. PGHS-2 mRNA abundance increases in amnion with labor onset, and this expression is correlated with increasing enzyme activity as indicated by Hirst et al. (1995). There is no change in PGHS-1 expression with labor onset. The factors at term which control the expression of PGHS-2 remain to be identified, but interleukin-1β, epidermal growth factor, renin, and glucocorticoids have all been shown to increase PGHS-2 mRNA in cultured amnion cells.

VI. LOSS OF INHIBITION IS ANOTHER MECHANISM FOR INITIATION OF CONTRACTION

Other physiological regulators of the PG biosynthetic pathway have been identified and postulated to play a role in the suppression of labor. It is hypothesized that withdrawal of these factors at parturition results in increased PG synthesis and uterine contractility.

A. Phospholipase A_2 Inhibitors

Lipocortins are PLA_2 inhibitors, but their physiological activity is questionable. They are found in large quantities which argues against a regulatory role. They appear to interact with the substrate rather than the enzyme which means they may not be specific for the PG biosynthetic pathway. Furthermore, inducible levels have not been found to correlate with phospholipase inhibition or PG synthesis. Uteroglobulin is a progesterone-induced protein inhibitor of PLA_2, and, although no physiological role for uteroglobulin has been established yet, it may be involved in the mediation of progesterone-controlled uterine quiescence.

Gravidin is a purported protein inhibitor of PLA_2 which is present in human amniotic fluid where it is identical to the secretory component of IgA. The greatest production of the gravidin protein has been found in the chorion. Assays in chorion taken before and after labor onset showed that gravidin activity before labor onset was much greater before labor onset than it was after. Thus, it is possible that gravidin becomes inactive or its synthesis stops at parturition. Measurements of gravidin have also been made in amniotic fluid and serum in a group of patients admitted in preterm labor. Mothers going into preterm labor had lower levels of gravidin than those who went on to term. In addition, serum gravidin-IgA falls to non-pregnant levels on administration of RU486. It therefore appears that gravidin action may be mediated by progesterone. These data provide evidence of a role for gravidin in the maintenance of pregnancy (Wilson, 1993). More recent work has disputed the existence of gravidin as a unique protein and as an inhibitor of PLA_2. Rather, it is believed that a different 72 kDa protein with PLA_2 inhibitory activity is present in human amniotic fluid.

B. PGHS Inhibitors

Human pregnancy plasma has been found to have the ability to inhibit PG synthesis. This has been ascribed to the presence of a circulating endogenous inhibitor of PG synthesis (EIPS) which inhibits cyclooxygenase. The physiological significance of this in the initiation of parturition is doubtful as EIPS levels did not change in studies of maternal plasma during pregnancy, parturition, and labor. However, it may act at a local level. Further, a pregnancy-associated prostaglandin synthase inhibitor (PAPSI) has been located specifically in human amniotic epithelial cells in women before the onset of labor and found to be absent in the amniotic epithelium of women in labor. The results from such studies may become more meaningful when reevaluated with reference to the recent findings concerning the differential expression of the PGHS -1 and -2 isoforms with parturition.

C. The Role of Nitric Oxide

In the rat uterus an L-arginine-nitric oxide-relaxation pathway exists that modulates contractility. The substrate, arginine, a donor of nitric oxide sodium nitoprusside and nitric oxide gas all produced substantial relaxation of the pregnant rat uterus. The effects of L-arginine were reversed by inhibitors of nitric oxide synthase (e.g., L-N^G-Nitroarginine methyl ester; L-NAME) and soluble guanylate cyclase, the receptor for nitric oxide. Nitrates and nitrites were produced by the uterus in the presence of L-arginine, and they were inhibited by L-NAME, indicating the involvement of nitric oxide synthase in the generation of nitric oxide. Furthermore there appeared to be a reduction in nitric oxide synthase activity at the time of delivery so the relaxation effects of L-arginine were decreased at the time of delivery. This was confirmed by detecting nitric oxide synthase (NOS) activity; markedly less NOS activity was detected in nerves, blood vessels, and decidua in uterine tissue collected during labor compared to the pregnant state. This indicates the L-arginine-nitric oxide system may contribute to the maintenance of uterine quiescence during pregnancy. A decrease in the generation of nitric oxide and/or in responsiveness to nitric oxide at term could be a further contributory mechanism that leads to increased uterine contractility and initiation of labor (Izumi et al., 1993). Indeed, recent work has indicated that in sheep myometrium a large increase in soluble guanylate cyclase occurs during the third trimester which then falls precipitously, implying that uterine quiescence may be due to changes in the nitric oxide receptor and not in nitric oxide levels. However, administration of L-NAME or nitroglycerin had no effect either shortening or lengthening gestation indicating that the overall cause of labor is not regulated by nitric oxide (see Garfield et al., 1995 for review).

In an experimental model, L-arginine inhibited contractions in human myometrial strips during the late stages of gestation suggesting a role for this system in human pregnancy. Based on the findings that nitric oxide activated the cyclooxygenase enzyme in the hypothalamus to produce PGE_2, it was shown that nitric oxide also stimulated the release of PGE_2, $PGF_{2\alpha}$, 6-keto $PFG1_{\alpha}$, TXB_2, and 5-HETE from 17β-estradiol treated rat uteruses *in vitro*. Clearly, this system must be studied under hormonal environments that have been otherwise manipulated, for example, under progesterone dominance or in the ovariectomized rat before its relevance in the control of pregnancy and parturition can be determined (Franchi et al., 1994).

VII. OTHER PHYSIOLOGICAL EVENTS OF PARTURITION

The essential aspect of successful delivery of the fetus is not only the development of rhythmical, sustained, and coordinated contractions of uterine muscle. Rupture of the fetal membranes and ripening and dilatation of the cervix must occur so that the fetus can be released from the uterine compartment.

A. Fetal Membrane Rupture

The process of labor and delivery results initially from a change in the relationship of the external fetal membranes and uterine wall. This involves separation of the chorion from the uterine decidual layer. There is enhanced glycosylation of a fetal-fibronectin glycoprotein localized in an area where the placenta and its membranes meet the uterine wall. The enhanced glycosylation of this placental fibronectin substantially reduces its binding affinity for other components of the extracellular matrix and therefore facilitates the separation of the chorion from the decidual layer. When this occurs fetal fibronectin is released into the cervicovaginal secretions. As fetal fibronectin is different to the fibronectin extracted from adult tissues, its appearance in these secretions is frequently used as a predictor of the onset of preterm and term labor and delivery (Ahner et al., 1995). Furthermore, there is evidence that separation of the chorion from the decidua may involve programmed cell death or apoptosis at the chorio-decidual interface. In rabbit uterine epithelial cells progesterone withdrawal and the presence of transforming growth factor β-1 has been shown to result in apoptosis. Both these events occur in the intrauterine tissues and are associated with the normal onset of parturition in this species (Gerschenson and Rotello, 1992).

Rupture of the fetal membranes is the result of mechanical and enzymatic processes. The biomechanical properties of fetal membranes collected after spontaneous labor or after cesarean section in the absence of labor are different. Following vaginal delivery the strength of the amnion decreases and the extensibility of the chorion increases. Thus, during labor there are mechanical changes in the fetal membranes that facilitate rupture. Other evidence suggests collagenases and fibrinolytic factors are involved in this process. Term amniotic fluids are capable of inducing the synthesis of collagenases and other proteases in fibroblasts while nonterm amniotic fluids fail to do the same. A number of fibrinolytic activators and inhibitors are present in the fetal membranes during gestation. It is postulated that a balance favoring the production of activators over inhibitors triggers membrane rupture at term (Watanabe et al., 1993).

B. Cervical Ripening

Further to membrane rupture, cervical ripening toward term and dilatation at delivery are essential processes. In the first part of gestation the cervix is hard and firmly holds the uterine contents. The biochemical process of "cervical maturation" commences at about the thirty-fourth week of pregnancy until the cervical os is fully dilated at delivery. The close cooperation between the myometrium and cervix is essential for normal uterine function, and defects in this relationship cause maternal and fetal morbidity.

There are three main structural components in the cervix of women: smooth muscle, collagen, and the connective tissue "ground-substance." The last contains the cervical glucosaminoglycans: dermatan sulfate, chondroitin sulfates, and hyaluronic acid. In humans, smooth muscle has not been shown to have a role in cervical dilatation (Uldjerg and Malmstrom, 1991). The enzymatic breakdown of collagen is a key factor in cervical softening. The collagen fragments become soluble and leave the ripened cervical tissue. The degradation of collagen occurs as an action of the enzymes collagenase and leukocyte elastase. The latter is located in the azurophil granules of polymorphonuclear leukocytes. Leukocyte infiltration and degranulation occurs in the term cervix in a similar manner to that seen in inflammatory reactions (Jeffery, 1991). Indeed, it was Liggins who first proposed that cervical ripening was similar to an inflammatory reaction. Cervical dermatan sulfate concentrations diminish along with those of collagen and the cervix becomes swollen and soft due to increased hyaluronic acid and water content. The increased hyaluronic acid and water content accounts for the soft, fragile texture of the ripened cervix, whereas the breakdown and loss of collagen and dermatan/chondroitin sulfates facilitate flexibility and distensibility.

The biochemical events underlying cervical maturation indicate it is an active cellular process and is thus subject to regulatory control. The activity of collagenase and other proteolytic enzymes rises with the increasing intrauterine estrogen dominance in late gestation. Conversely, in non-pregnant human cervix explants, collagen breakdown is diminished by progesterone administration. PGs, especially PGE_2, are clearly involved in cervical ripening at term in women. PGs have been used clinically for years to induce first and second trimester abortions and cervical ripening. Further, in humans ripening of the cervix is associated with increased PGI_2 and HETE production. The latter are arachidonic acid metabolites produced by lipoxygenase enzymes. HETEs and their metabolites are potent chemoattractants; PGI_2 is involved in increasing vascular permeability during inflammatory reactions. Based on these observations it is possible production of these mediators summons the polymorphonuclear leukocytes known to infiltrate the cervix at term and leads to enzyme secretion and collagen degradation.

In addition to evidence of their local production by the cervix, PGs have been shown to have effects on cervical ripening *in vivo*. In late pregnant sheep treated with epostane, a 3βHSD inhibitor which decreases progesterone synthesis, there was an increase in utero-ovarian plasma PGE_2 and $PGF_{2\alpha}$. This was accompanied by increases in uterine activity and cervical softening. Addition of the PG synthesis inhibitor, mefenamic acid, caused PG levels to fall and uterine activity and cervical softening to cease.

The hormone relaxin is also postulated to be involved in cervical ripening. In non-pregnant, estrogen-primed rhesus monkeys, relaxin administration induced histologic, biochemical, and biomechanical changes that were similar to normal cervical ripening. Relaxin receptors are present in the human cervix and local administration of relaxin to women is beneficial in cervical dilatation.

Last, a relationship between cervical maturation and the initiation of labor is well demonstrated. In one study women received labor preinduction treatment of oxytocin infusion alone or oxytocin infusion and intracervical PGE_2 gel. The contractile activity was no different in the two groups. However, in the PGE_2 treated group, the women proceeded to spontaneous labor and delivered fast; the length of the active phase and the second stage of labor was shorter, and the incidence of cesarean sections was lower. Thus, pretreatment of the cervix with PG suppositories causing maturation facilitates a more efficient labor process without an increase in myometrial activity (for a review see Huszar and Walsh, 1991).

VIII. SUMMARY

It is evident that parturition results from a complex interplay of maternal and fetal factors. It is the result of the sequential maturation of an endocrine organ communication system. In sheep, and potentially primates, the sequence can be seen to begin at the level of the fetal brain with increased cortisol production from the fetal adrenal providing the trigger to the subsequent evolution of maternal endocrine changes. In sheep, the regulatory mechanisms responsible for the changes in fetal adrenal glucocorticoid secretion can be considered to be the initial signals of the onset of parturition. In primates, these mechanisms are estrogen-dependent, the synthesis of which is directed by the arrival of C_{19} steroid precursors at the placenta. These estrogen precursors are also synthesized by the fetal adrenal. Throughout gestation, progesterone acts to maintain pregnancy. Interestingly, in primates placental progesterone synthesis is for the most part dependent on estrogen synthesis by the same tissue and therefore, ultimately, the fetus. Toward term, the influence of progesterone is either withdrawn locally and/or overcome by rising estrogen levels. The role of the fetal adrenal in providing increased estrogen precursors for the latter process in primate parturition is at least firmly established. Under the increasing estrogen dominance of the intrauterine tissues, activation of the myometrium and an increased capacity for uterotonin synthesis occurs. This leads to the rhythmic contractions of labor that lead to the expulsion of the uterine contents. Rupture of fetal membranes and dilatation of a ripened cervix are essential to complete a successful delivery. A glucocorticoid signal from the fetus, or, another as yet undefined signal, initiates a feed-forward cascade of events which results in secretion of myometrial stimulants, uterine contraction, membrane rupture, cervical dilatation, and labor. The ultimate signals for birth may yet be found to reside at the level of the primate fetal brain as in the sheep. These signals initiate an array of fetal and maternal endocrine and paracrine processes that result in the timely delivery of a viable neonate.

ACKNOWLEDGMENTS

Studies in the authors' laboratory were supported by the Medical Research Council of Canada, the Alberta Heritage Foundation for Medical Research, the University of Alberta Perinatal Research Centre, and the University of Alberta Hospitals.

REFERENCES

Ahner, R., Kiss, H., Egarter, C., Zeillinger, R., Eppel, W., Karas, H., & Husslein, P. (1995). Fetal fibronectin as a marker to predict the onset of term labor and delivery. Am. J. Obstet. Gynecol. 172, 134–137.

Alexandrova, M., & Soloff, M.S. (1980). Oxytocin receptors and parturition I. Control of oxytocin receptor concentration in the rat myometrium at term. Endocrinology 106, 730–735.

Burghardt, R.C., Barhoumi, R., & Dookwah, H. (1993). Endocrine regulation of myometrial gap junctions and their role in parturition. Semin. Reprod. Biol. 11(3), 250–260.

Casey, M.L., MacDonald, P.C., & Mitchell, M.D. (1985). Despite a massive increase in cortisol secretion in women during parturition, there is an equally massive increase in prostaglandin synthesis. A paradox. J. Clin. Invest. 75, 1852–1857.

Challis, J.R.G., Lye, S.J., Mitchell, B.F.M., Olson, D.M., Sprague, C., Norman, L., Power, S.G.A., Siddiqi, J., & Wlodek, M.E. (1985). Fetal signals for birth. In: *The Physiological Development of the Fetus and Newborn* (Jones, C.T., & Nathanielsz, P.W., Eds.) pp. 363–369. Academic Press, New York.

Challis, J.R.G., & Olson, D.M. (1988). Parturition. In: *The Physiology of Reproduction* (Knobil, E., Neill, J., Ewing, L.L., Greenwald, G.S., Markert, C.L., & Pfaff, D.W., Eds.) pp. 2177–2216. Raven Press Ltd., New York.

Csapo, A.I. (1976). The "see-saw" theory of parturition. In: *The Fetus and Birth,* (Knight, J., & O'Connor, M.O., Eds.). p. 159. Ciba Foundation Symposium No. 47, Elsevier, Amsterdam.

Franchi, A.M., Chaud, M., Rettori, V., Suburo, A., McCann, S.M., & Gimena, M. (1994). Role of nitric oxide in eicosanoid synthesis and uterine motility in estrogen treated rat uteri. Proc. Natl. Acad. Sci. USA 91, 539–543.

Fuchs, A.P. (1995). Plasma membrane receptors regulating myometrial contractility and their hormonal modulation. Semin. Perinatol. 19(1), 15–30.

Garfield, R.E., Ali, M., Yallampalli, C., & Izumi, H. (1995). Role of gap junctions and nitric oxide in control of myometrial contractility. Semin. Perinatol. 19(1), 41–51.

Gerschenson, L.E., & Rotello, R.J. (1992). Apoptosis: A different type of cell death. FASEB J. 6, 2450–2455.

Hansen, H.S. (1976). 15-Hydroxy PG dehydrogenase. A review. Prostaglandins 12, 647–649.

Hirst, J.J., Chibba, R., & Mitchell, B.F. (1993). Role of oxytocin in the regulation of uterine activity during pregnancy and in the initiation of labor. Semin. Reprod. Endocrinol. 11(3), 219–233.

Hirst, J.J., Teixeira, F.J., Zakar, T., & Olson, D.M. (1995). Prostaglandin endoperoxide-H synthase-1 and 2 mRNA levels in human amnion with spontaneous labor onset. J. Clin. Endocrinol. Metab. 80, 517–523.

Holm, L.W. (1967). Prolonged pregnancy. Adv. Vet. Sci. Comp. Med. 11, 159.

How, H., Zuo, J., & Lei, Z.M. (1994). Human myometrial estradiol & progesterone receptor changes in preterm and term deliveries. Abstract. 41st Annual Meeting of the Society for Gynecologic Investigation, Chigaco, IL, March 22–26.

Huszar, G.B., & Walsh, M.P. (1991). Relationships between myometrial and cervical functions in pregnancy and labor. Semin. Perinatol. 15(2), 97–117.

Izumi, H., Yallampalli, C., & Garfield, R.E. (1993). Gestational changes in L-arginine induced relaxation of pregnant rat and human myometrial smooth muscle. Am. J. Obstet. Gynecol. 169(5), 1327–1337.

Jeffery, J.J. (1991). Collagen and collagenase: Pregnancy and parturition. Semin. Perinatol. 15(2), 118–126.

Kennedy, P.C. (1971). Interaction of fetal disease and the onset of labor in cattle and sheep. Fed. Proc. 30, 110.

Liggins, G.C., Fairclough, R.J., Grieves, S.A., Forster, C.S., & Knox, B.S. (1976). Parturition in the sheep. In: *The Fetus and Birth*, (Knight, J., & O'Connor, M.O., Eds.) p. 5. Ciba Foundation Symposium No. 47, Elsevier, Amsterdam.

Lye, S. (1994). The initiation and inhibition of labor—Toward a molecular understanding. Semin. Reprod. Endocrinol. 12(4), 254–294.

Malpas, P. (1933). Postmaturity and malformation of the fetus. Br. J. Obstet. Gynaecol. 40, 1046.

Olson, D.M., Zakar, T., & Mitchell, B.F. (1993). Prostaglandin synthesis regulation by intrauterine tissues. In: *Molecular Aspects of Placental and Fetal Membrane Autocoids*, (Rice, G.E., & Brennecke, S.P., Eds.) pp. 55–95. CRC Press, Boca Raton, FL.

Olson, D.M., Mijovic, J.E., & Sadowsky, D. (1995). The control of human parturition. Semin. Perinatol. 19(1), 52–63.

Pepe, G.J., & Albrecht, E.D. (1990a). Placental steroid biosynthesis in primate pregnancy. Endocr. Rev. 11(1), 124–150.

Pepe, G.J., & Albrecht, E.D. (1990b). Regulation of the primate fetal adrenal cortex. Endocr. Rev. 11(1), 151–176.

Riley, S.C., & Challis, J.R.G. (1991). Corticotrophin-releasing hormone production by the placenta and fetal membranes. Placenta 12, 105–119.

Swaab, D.F., Boer, K., & Honnebier, W.J. (1976). Influence of the fetal hypothalamus and pituitary on the onset and course of parturition. In: *The Fetus and Birth* (Knight, J., & O'Connor, M.O., Eds.) p. 359. Ciba Foundation Symposium No. 47, Elsevier, Amsterdam.

Thorburn, G.D. (1994). A time to be born. Semin. Reprod. Endocrinol. 12(4), 213–226.

Uldjerg, N., & Malmstrom, A. (1991). The role of proteoglycans in cervical dilatation. Semin. Perinatol. 15(2), 127–132.

Watanabe, T., Araki, M., Mimuro, J., Tamada, T., & Sakata, Y. (1993). Fibrinolytic components in fetal membranes and amniotic fluid. Am. J. Obstet. Gynecol. 168(4), 1283–1289.

Wilson, T. (1993) Gravidin: An endogenous inhibitor of phospholipase A_2. Gen. Pharmac. 24 (6), 1311–1318.

Word, R.A. (1995). Myosin phosphorylation and the control of myometrial contraction/relaxation. Semin. Perinatol. 19 (1), 3–14.

THE TROPHOBLAST AS AN ACTIVE REGULATOR OF THE PREGNANCY ENVIRONMENT IN HEALTH AND DISEASE:

AN EMERGING CONCEPT

Donald W. Morrish, Jamal Dakour, and Hongshi Li

Advances in Organ Biology
Volume 1, pages 121–152.

ISBN: 1-55938-639-8

ABSTRACT

The trophoblast is increasingly recognized as an active regulator of its environment and no longer a passive steroid-producing, immunological filter. Considerable evidence supports a dominant role of the trophoblast in blastocyst implantation, maternal metabolic regulation, regulation of the maternal immune response to the allograft fetus, angiogenesis, and regulation of placental blood flow and parturition. These actions are accomplished largely through the secretion of soluble substances, but also involve the expression of unique surface antigens such as HLA-G. Currently, there is poor understanding of the pathophysiological mechanism of several diseases of pregnancy: intrauterine growth retardation, pregnancy-induced hypertension, and preterm labor. The evolving knowledge of trophoblast regulatory functions within the uterus promises opportunities for clarifying disease mechanisms.

I. INTRODUCTION

Concepts of placental function have evolved from its first recognition as a protein-producing organ (Halban, 1905) to a participant in a complex endocrine-paracrine system involving the mother, placenta, and fetus (Jones, 1993; Cross et al., 1994). What, then, are the functions or controlling factors within this system that can be regarded as primarily originating in the trophoblast? This chapter will seek to outline established dominant trophoblast functions and to consider emerging possible new leading roles of the trophoblast in normal physiology and in the pathophysiology of diseases of pregnancy.

II. IMPLANTATION

A. Physiology

The ability of the blastocyst to implant ectopically is perhaps the best evidence of the leading role of trophoblast cells in implantation, morphologically evident in very early embryo attachment at seven days after fertilization (Hertig and Rock, 1945). The ability of the trophoblast to attach and invade the uterus appears to have many similarities to the mechanism of invasion of metastasizing malignant cells (Liotta, 1990; Lala and Graham, 1990).

What are the properties of the trophoblast that permit this activity? Pre-implantation human blastocysts make β1 integrin and laminin in preparation for attachment (Turpeenniemi-Hujanen et al., 1992; Figure 1). After implantation, both the cytotrophoblast and the endometrium express cell adhesion molecules such as integrins (Korhonen et al., 1991; Damsky et al., 1992; Lessey et al., 1992) and E-cadherin (Glasser et al., 1988; Fisher et al., 1989) which are likely to be important in attachment of the blastocyst to the endometrium. These have not been studied in the fallopian tube but presumably must be analogously present to permit attachment in this location. The expression of these proteins represents an inherent property of the cytotrophoblast cells for attachment behavior. The cytotrophoblast also synthesizes an unique fibronectin isoform, oncofetal fibronectin, which may act as an additional "glue" to specifically attach the blastocyst onto endometrial integrin receptors (Feinberg et al., 1991).

The next step is release of proteases that degrade the endometrial stroma to permit entrance of the invading cytotrophoblast. In preparation for invasion, the human pre-implantation blastocyst synthesizes 92 kd type IV collagenase that increases up to the time of implantation (Puistola et al., 1989). Mouse blastocysts also synthesize other metalloproteinases, tissue inhibitors of metalloproteinases (TIMPs), and urinary-type plasminogen activator (uPA) but it is not known if the human blastocyst is similar (Strickland et al., 1976; Sappino et al., 1989; Behrendtsen et al., 1992; Brenner et al., 1989). Once implanted, the invasive extravillous cytotrophoblast releases predominantly 92 kD type IV collagenase (Fisher et al., 1985, 1989; Librach et al., 1991; Fernandez et al., 1992), but possibly other collagenases, aminopeptidases, cathepsin B (Gossrau et al., 1987; Liotta, 1992; Yagel et al., 1993; Shimonovitz et al., 1994;

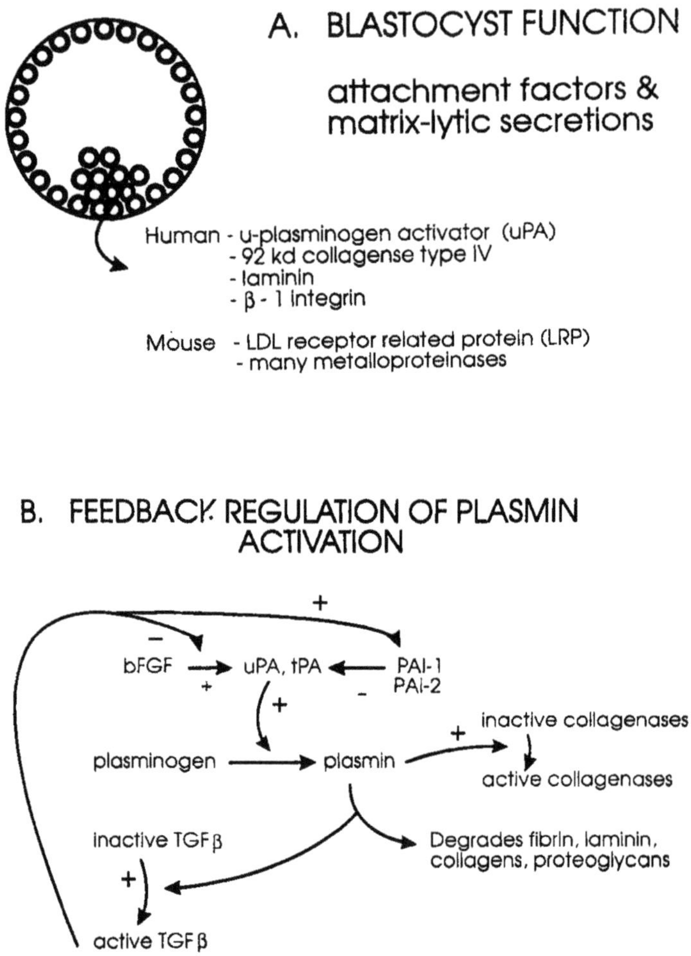

Figure 1. (A) Pre-attachment blastocyst trophectodermal cells express integrins for attachment, secrete lytic factors (uPA, collagenase) to aid invasion of the endometrial surface, and laminin for attachment and matrix support within the endometrium. (B) A complex regulatory system of matrix degradatory factors (uPA, plasmin, collagenases) is present at the trophoblast-endometrial interface.

Emonard et al., 1990; Autio-Harmainen et al., 1992), and uPA (Martin and Arias, 1982; Queenan et al., 1987). uPA converts plasminogen to plasmin, a serine protease with broad substrate specificity including collagens, laminin, and fibronectin (Vaheri et al., 1990; Figure 1). Invading cytotrophoblast cells express uPA receptor, thus making uPA available at the invasion site (Zini et al., 1992). Gene knockout experi-

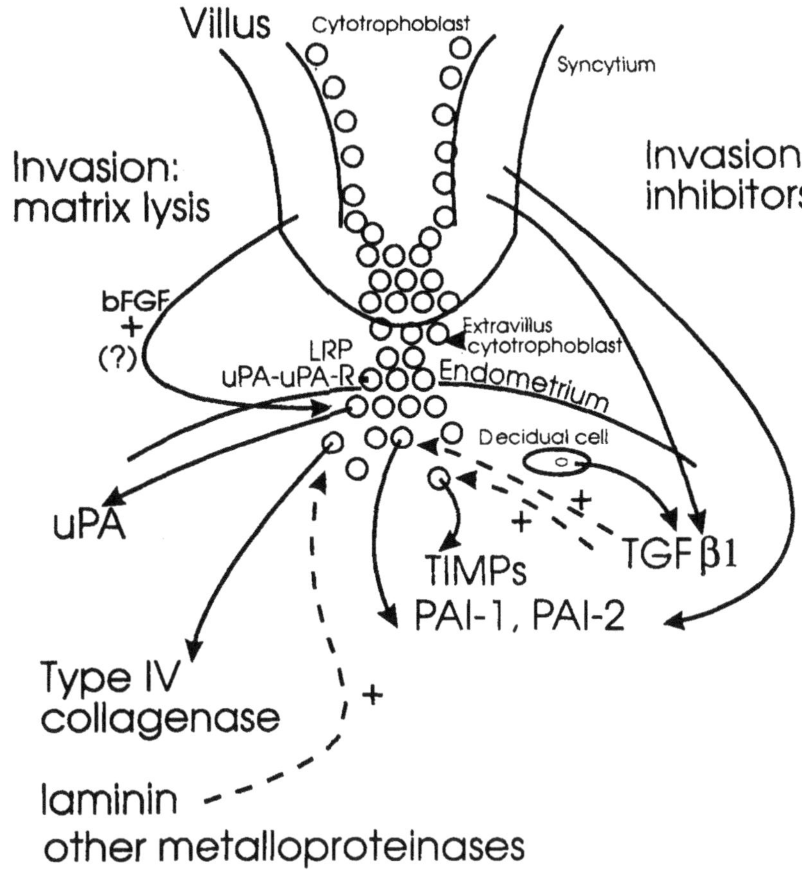

Figure 2. Implantation is a complex interactive process between the trophoblast and endometrium, driven primarily by trophoblast-derived factors. Invasion-promoting factors (left side of figure) include bFGF, LRP, uPA, and collagenases. There appears to be a positive feedback loop wherein trophoblast-secreted laminin increases type IV collagenase production. Invasion-inhibitors (right side of figure) include TIMPs, PAI-1, PAI-2 and TGFβ1 (which further induces TIMPs and PAI-1 and PAI-2). The trophoblast thus controls both stimulators and inhibitors in a balanced system.

ments in mice also point to a role for low density lipoprotein-related protein (LRP) in clearing uPA-urinary plasminogen activator-plasminogen activator inhibitor (uPA-PAI) complexes from the surface of invading cytotrophoblasts, thus making uPA receptor sites available for new active enzyme to be presented by the cytotrophoblasts to the endometrium (Herz et al., 1992). Interestingly, the cytotrophoblasts also make laminin (Autio-Harmainen et al., 1991) which stimulates collagenase IV activity (Emonard et al., 1990), thus providing a positive feedback loop to enhance invasion as well as provide a substrate for cytotrophoblast attachment. Although not yet shown to occur in cytotrophoblast cells, basic fibroblast growth factor (bFGF) stimulates uPA production in endothelial cells (Flaumenhaft et al., 1992). As bFGF is made predominantly in the syncytium (Cattini et al., 1991), this would provide yet another positive regulatory mechanism for activating degradative enzymes.

The trophoblast also produces a variety of degradation inhibitors to counterbalance these invasion-promoting enzymes. The trophoblast synthesizes plasminogen activator inhibitors (PAI-1, PAI-2; Feinberg et al., 1989), and tissue inhibitors of metalloproteinases (TIMPs), the major collagenase inhibitors in placenta (Lala and Graham, 1990; Rajabi et al., 1990; Graham and Lala, 1992). The TIMPs can inhibit cytotrophoblast invasion analogous to their effect on tumor invasion (Lala and Graham, 1990; Testa and Quigly, 1991). Additionally, the trophoblast, and particularly the syncytium (Figure 3) as well as decidua can make TGFβ1, which can induce the TIMPs. It is possible that plasmin, ubiquitous in blood, may activate latent TGFβ1 produced in trophoblast and decidua (Lyons et al., 1988; Matsuzaki et al., 1992; Graham and Lala, 1992). TGFβ1, in turn, could inhibit uPA activity as occurs in endothelial cells (Flaumenhaft et al., 1992; Grainger et al., 1995). The syncytium also expresses much more PAI-2 (Feinberg et al., 1989) and appears to express relatively more PAI-1 (Figure 3; Morrish et al., unpublished data).

The cytotrophoblast thus has the capability of regulating its own invasion (Figures 1 and 2) by controlling the secretion of both degrading enzymes, their inhibitors, and some of their regulators.

After degradation of the endometrial extracellular matrix, the cytotrophoblast probably lays down matrix proteins favorable to its own growth. There is evidence that the cytotrophoblast secretes fibronectin (both oncofetal and mature isoforms; Queenan et al., 1987; Feinberg et al., 1991; Morrish et al., unpublished data) and localization studies have shown the mRNA and peptide for laminin and collagen type IV in cytotrophoblast columns (Autio-Harmainen et al., 1991). Collagen I mRNA is also expressed in cytotrophoblast cells (Figure 3; Morrish et al., unpublished data.)

Since the cytotrophoblast initiates invasion, it follows that decidual-produced modifiers of the process such as extracellular matrix binding

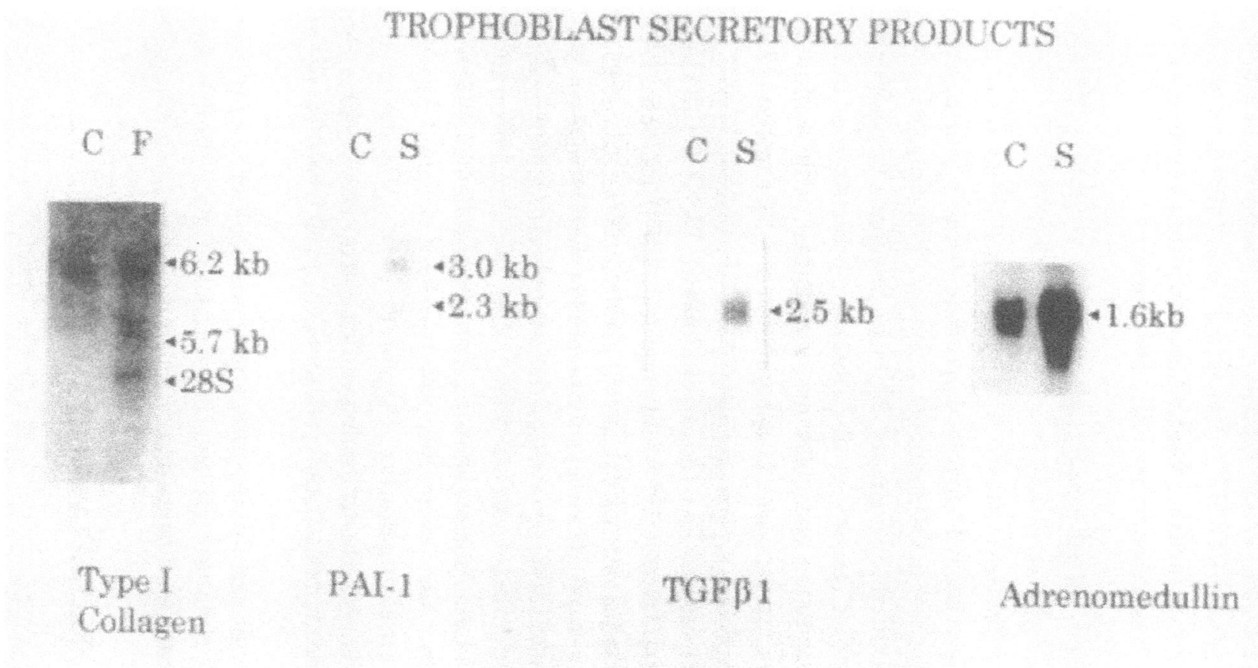

Figure 3. Northern blots of trophoblast products using *in vitro* cultured human term trophoblast cells as described (Morrish et al., 1987). 15 μg total cellular mRNA was transferred to nylon membranes and hybridized with specific human cDNA probes as shown. (A) human collagen type I in cytotrophoblast (C) and placental fibroblasts (F); (B) PAI-1 expression in cytotrophoblast (C) and partially-formed syncytium *in vitro* (S); (C) TGFβ1 in cytotrophoblast (C) and syncytium (S); (D) adrenomedullin expression in cytotrophoblast (C) and syncytium (S).

via integrins (Werb et al., 1990; Hynes, 1992) or insulin-like growth factor binding protein (IGFBP-1; Irving and Lala, 1994) may serve as feedback signals by which the cytotrophoblast can direct synthesis of the appropriate matrix elements to suit its environment.

B. Pathophysiology

Intriguing evidence exists that a primary abnormality in some of these regulatory processes may be involved in certain diseases, although there is as yet no proof that the abnormalities originate in the placenta. Some types of female infertility are associated with disruption of endometrial integrin expression (Lessey et al., 1992). In preeclampsia, cytotrophoblast differentiation and invasion are abnormal, with shallow endometrial penetration and lack of upregulation of $\alpha 1/\beta 1$ integrin (Zhou et al., 1993). Preeclamptic patients have decreased placental PAI-2 and uPA levels and increased PAI-1 levels (Reith et al., 1993; Estelles et al., 1994; Lindhoff and Astedt, 1994; Hofmann et al., 1994). Increased PAI-2 levels are seen in patients with recurrent abortion (Gris et al., 1993). In both diseases, decreased fibrinolysis is favored by these changes in PAI-2 or PAI-1, leading to fibrin deposition that may block trophoblast development or interfere with vascularization, thus promoting poor attachment. Although initial studies showing high cervical tissue collagenase levels in preterm labor were speculated to be a potential placental disorder leading to early parturition (Rajabi et al., 1987), this abnormality appears more likely due to neutrophils infiltrating the cervix and not due to increased placental production (Osmers et al., 1992; Morrison et al., 1994).

Thus, several diseases of pregnancy may potentially have a primary abnormality in trophoblast implantation as a cause. Clearly, further study is needed to determine a cause and effect relationship.

III. METABOLIC REGULATION

Interestingly, and perhaps teleologically appropriately, the syncytium, the transporter of nutrients to the fetus, is the source of the major metabolically active hormones.

A. hCG

The longest recognized regulatory function of the placenta is the secretion of human chorionic gonadotropin (hCG), first discovered in 1927 (Aschheim and Zondek, 1927) and shown to originate in the placenta by Kido (1937). HCG functions in the first trimester to stimulate corpus luteum estrogen and progesterone production, both essential for maintaining the endometrium and for early blastocyst development (Falcone and Little, 1994; Figure 4). The syncytium is the predominant source of both α and β subunits of hCG, but a few villous, cell column, and extravillous cytotrophoblast cells also appear to make these subunits (Hoshina et al., 1982, 1983; Kurman et al., 1984; Sasagawa et al., 1987). Other roles for hCG may include stimulation of fetal testosterone synthesis and Leydig cell differentiation (Huhtaniemi et al., 1977; Pelliniemi

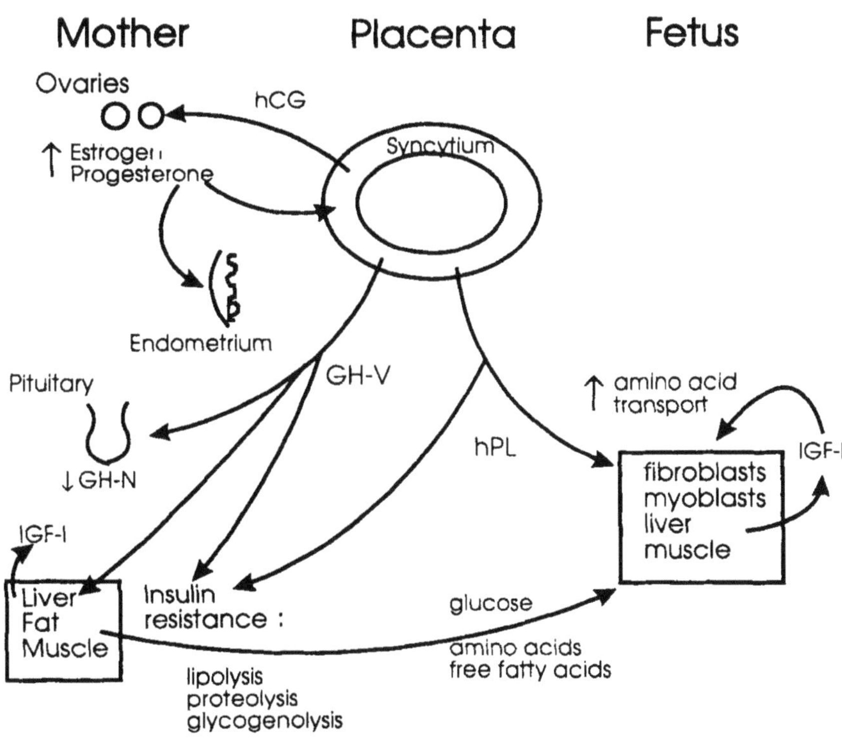

Figure 4. The syncytium produces several factors, hCG, hPL, and GH-V, that regulate maternal metabolism and that may regulate fetal growth and metabolism as well. (Modified from D.J. Hill, 1992, with permission).

and Dym, 1980), placental glucose production (Cedard et al., 1970), and fetal dehydroepiandrosterone sulfate (DHEAS) production (Seron-Ferre et al., 1978).

B. HPL (Human Placental Lactogen)

Human placental lactogen is a member of the growth hormone (GH) gene family, having 91–99% homology in the coding region with GH (Chen et al., 1989). Synctium appears to be the sole source of hPL (McWilliams and Boime, 1980; Tabarelli et al., 1983). HPL has major metabolic effects on the mother causing insulin resistance and leading to lipolysis with increased free fatty acids, amino acid mobilization, and glycogenolysis (reviewed in Hill, 1992; Jones, 1993; Falcone and Little, 1994). The purpose of these actions is presumably to make substrate available to the fetus. HPL may also have direct growth-promoting effects on fetal cells (reviewed in Hill, 1992; Figure 4).

C. GH-V (Growth Hormone Variant)

The nucleotide sequence for GH predicted a new gene, GH-V (Seeburg, 1989), and it has since been found in the syncytium of the placenta (Frankenne et al., 1987). During gestation, maternal GH-V levels rise and pituitary GH (GH-N) decreases so that GH-V constitutes nearly all the GH activity by term (Hennen et al., 1987). Thus, GH-V replaces maternal GH in regulating maternal metabolism, with similar effects to those of hPL on glucose and lipid metabolism (Figure 4).

D. Pathophysiology

Without hCG to maintain steroid production, a pregnancy fails. However, curiously, there are no described diseases of "partial hCG" deficiency, or abnormal structural hCG leading to miscarriages. An inadequate maternal luteal phase usually has been blamed for early miscarriages, but therapy of recurrent abortion with progesterone has not convincingly proven effective (Hill, 1994). HPL levels are lower in intrauterine growth retardation, maternal hypertension, preeclampsia, and threatened abortion, but there are no data to support decreased hPL as a primary cause of these conditions (Kelly et al., 1975; Spellacy et al., 1974). GH-V has not been studied in any disease condition. Although several metabolic effects of hPL and GH-V exist (Hill, 1992), the true role of hPL and GH-V is unclear, as women with gene deletions have normal pregnancies, placentas, and fetuses (Simon et al., 1986). Clearly, there are potential undiscovered abnormalities in these hormones that may cause disease.

IV. IMMUNE MODULATION

The antigenically-foreign fetus escapes rejection by the maternal immune system. The simple concept of the placenta as an immunological barrier is clearly false as the trophoblast is in intimate maternal contact where it is attached to the decidua and trophoblast cells can be found in the maternal circulation (Jaameri et al., 1965; Goodfellow and Taylor, 1982). In recent years, it has become apparent that, although immunological tolerance is clearly a complex process involving both maternal and placental factors (reviewed in Sargent et al., 1994; Starkey, 1994), three key elements in the puzzle are trophoblast expression of particular antigens, immunosuppressive factors, and cytokines (Figure 5).

HLA-G, a novel class of type I MHC antigen, is found only in the placenta, and expression of the protein occurs only in extravillous cytotrophoblast cells (Ellis, 1990; Kovats et al., 1990). In contrast, classical MHC class I and II antigens are expressed only in the villous stromal core. Although the mechanism of action in preventing fetal rejection is unclear, several possibilities have been proposed. HLA-G may inhibit natural killer (NK) cell activity (Ferry et al., 1991), because the NK cells (in the decidua, the large granular lymphocytes, LGL) do not act against cells expressing HLA-I. HLA-G may activate maternal suppressor cells in the decidua (Starkey et al., 1988). It has also been suggested HLA-G might bind to receptors of cytotoxic T cells and block recognition of non-MHC target structures on trophoblasts, or that, being non-polymorphic, HLA-G-expressing cells would not stimulate MHC-restricted rejection (Kovats et al., 1990). Interestingly, cytotrophoblast and syncytium also do not stimulate a cytotoxic T cell response (Hunt et al., 1984; Khalfoun et al., 1986; Sargent and Redman, 1989). The mechanism of this evasion of attack is unknown.

Two other antigens expressed by trophoblast may modulate local action of the complement lysis system. CD46 (formerly called TLX, trophoblast/lymphocyte crossreactive antigen) is known to bind the C3b and C4b components of complement, thus blocking activation of this pathway (Purcell et al., 1990). CD55 ("decay accelerating factor" or DAF) also regulates C3 convertase enzymes and is present on tro-

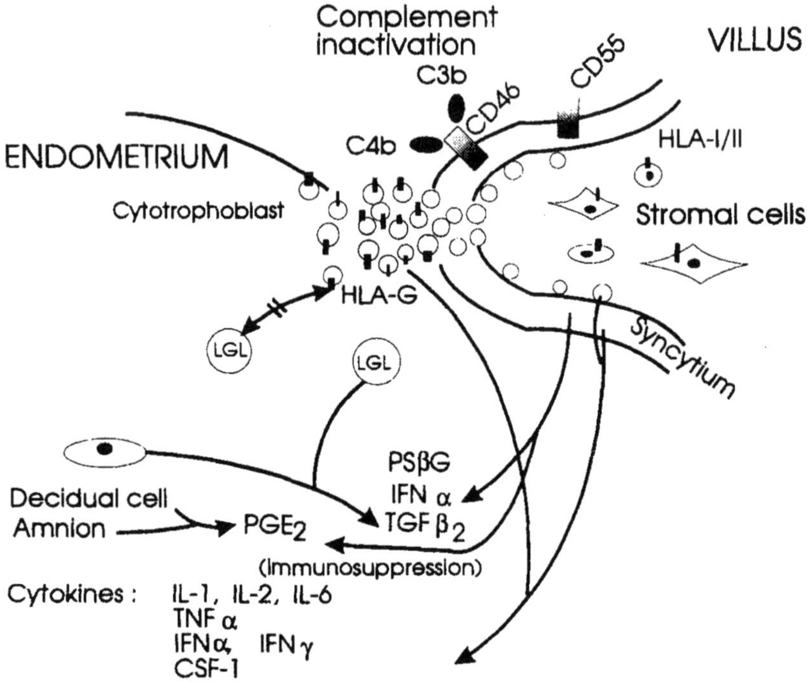

Figure 5. The trophoblast may modulate maternal immune function to prevent rejection of the fetal allograft by several mechanisms. Extravillous cytotrophoblast expresses HLA-G, which may induce tolerance in maternal large granular lymphocytes (LGL) with NK activity or induce other favorable immune responses. Syncytium also expresses CD46 and CD55, which may serve to inactivate any complement-mediated lysis directed at the trophoblast. Trophoblast, LGLs, and decidual cells all secrete TGFβ_2 which appears to have immunosuppressant activity. PGE$_2$ originates from trophoblast, decidua, and amnion and is long known to be immunosuppressant. The syncytial factors IFNα and pregnancy specific β glycoprotein (PSβG) also inhibit T cell function. Many trophoblast-derived cytokines have pleiotropic functions, some of which promote maternal immunosuppression.

phoblast (Hunt and Hsi, 1990). These would further inhibit maternal immune action against the placenta.

The second manner in which the trophoblast appears to inhibit the maternal immune response is by the secretion of soluble immunosuppressants. Both decidual cells (presumably the large granular lymphocytes) and syncytium secrete a substance closely-related to, or identical with, TGFβ2 which has been identified as at least one of the immunosuppressant factors (Altman et al., 1990; Clark et al., 1990; Starkey, 1994; Lea et al., 1990, 1992). It has long been recognized that prostaglandin E2 (PGE_2) *in vitro* is a potent immunosuppressant (Goldyne and Stobo, 1981; Weir et al., 1991). PGE_2 is produced by trophoblast, amnion, and decidua (see Section VI). As well, two syncytium-derived products, interferon α (IFNα) and pregnancy-specific β glycoprotein (SP1), have immunosuppressant activity (Harris et al., 1984; Bulmer et al., 1990; Lea and Clark, 1991).

The trophoblast also produces many cytokines including interleukins (IL-1, IL-2, IL-6), colony stimulating factor 1 (CSF-1; Daiter et al.,1992), and interferons (IFNα, IFNγ; Starkey, 1994). The possible relationships with the maternal immune system are exceedingly complex, involving non-immune effects of the cytokines and are beyond the scope of this article (reviewed in Starkey, 1994; Wegmann et al., 1993). The interactions are made more complex by the production of many of these cytokines and their receptors by decidual cells and placental villus macrophages as well as trophoblast, indicating paracrine and autocrine control loops. Possible functions of these cytokines include regulation of other cytokine production (IL-1, IL-6, M-CSF, GM-CSF), activation of T cells, modulation of trophoblast proliferation, differentiation, and peptide hormone production (Garcia-Lloret et al., 1994) and stimulation of LGL and macrophage cytotoxicity (reviewed in Starkey, 1994). The precise interaction and balancing of competing influences of these cytokines is unclear but serves to demonstrate the potent and active role of the trophoblast.

A. Pathophysiology

No data currently exist demonstrating abnormalities in HLA-G, CD46, or CD55 expression, IFNα, PGE_2 or TGFβ2 in any human disease state. There are also no reports of abnormal serum cytokine levels in IUGR or preeclampsia. However, mitogen-stimulated IFNγ production from monocytes is lower in neonates with IUGR or preterm delivery

(Saito et al., 1992). The involvement of the immune system in recurrent abortion remains an intriguing and controversial topic (reviewed in Hill, 1994). Clearly, abnormalities in any component of this system may lead to a miscarriage. Several studies show abnormal cytokine production in preterm labor and these will be discussed subsequently. Abnormal cytoregulation deserves further study as an etiological factor in IUGR and preeclampsia.

V. REGULATION OF ANGIOGENESIS AND BLOOD FLOW

A. Angiogenesis

Since the trophoblast is the first differentiated cell type to form in the blastocyst, it has the potential of being able to regulate formation of other cells and structures. Recent data suggests that regulation of angiogenesis may be one such function. Stimulators of angiogenesis include acidic FGF (aFGF; FGF-1), bFGF (FGF-2), platelet-derived endothelial cell growth factor (PDECGF), vascular endothelial growth factor (VEGF), angiogenin, endothelin 1 (ET-1), and others. Inhibitors of angiogenesis include thrombospondin, angiostatin, and other less well-defined molecules (Folkman, 1995), but there are no reports of any of these being found in the placenta. Basic FGF, one of a seven-member FGF peptide family (Baird, 1993), was originally purified from the placenta (Gospodarowicz et al., 1985) and has been localized by immunoperoxidase staining principally to the syncytium (Cattini et al., 1991). The FGFs are generally the most potent angiogenic substances known. This multifunctional growth factor may also have the unexpected effect of stimulating hCG secretion (Oberauer et al., 1988). ET-1, a potent vasoconstrictor (discussed further following) is also mitogenic to smooth muscle cells and vessel fibroblasts (Bonin et al., 1993). EGF and bFGF increased the number of ET-1 binding sites and were thus synergistic for mitogenesis in these studies.

PDECGF and VEGF appear to have unique placental isoforms (Usuki et al., 1990). Jackson et al. (1994) have studied distribution of these two factors in placenta. PDECGF was found in the syncytium, villus stroma, and endothelium in first trimester placentas, but staining intensity in the trophoblast declined during gestation. VEGF, in contrast, was almost exclusively found in the trophoblast, being predominantly in the cytotrophoblast in the first trimester, and in the syncytium later in gestation.

Similarly, VEGF is localized to epithelial cells and myocytes but not to endothelium in a variety of fetal tissues, suggesting a general paracrine regulation of angiogenesis (Shifren et al., 1994). Data in rats show a dominant trophoblast localization of VEGF in early embryos, with other tissues becoming significant production sites only later in gestation (Jakeman et al., 1993). Together, these results support the concept that the trophoblast is dominant in the generation of the placental and probably the fetal vasculature (Figure 6).

B. Blood Flow

Since the placenta lacks innervation (Spivack, 1943; Walker and MacLean, 1971), humoral factors must regulate vascular resistance. A very large number of factors have been studied and, along with concepts of their action, have been excellently reviewed (Boura and Walters, 1991; Myatt, 1992; Walters, 1992). Factors causing constriction are oxygen, endothelins 1 and 3 (ET-1, ET-3), angiotensin II, kinins, catecholamines, leukotrienes (LTC_4, LTD_4), thromboxane A_2, 5-hydroxytryptamine (5-HT), neuropeptide Y (NPY), and arginine vasopressin (AVP). Vasodilators include nitric oxide, prostacylin (PGI_2), calcitonin gene-related peptide (CGRP), histamine, substance P, vasoactive intestinal peptide (VIP), and atrial natriuretic peptide (ANP; Myatt, 1992). Another potential vasodilator is bFGF (Cuevas et al., 1991). A newly discovered vasodilator, adrenomedullin, also is present in placenta (Kitamura et al., 1993a, 1993b; Morrish et al., unpublished data; Figure 5). There are no reports demonstrating trophoblast production of catecholamines, AVP, VIP, CGRP, substance P, or ANP (although, interestingly, ANP mRNA is present in human umbilical vessels; Cai et al., 1993).

However, several potent factors are synthesized by the trophoblast including ET-1, prostaglandins, NPY, nitric oxide, bFGF, and adrenomedullin. bFGF has been alluded to earlier. Adenosine and other purines are ubiquitous cellular products that produce vasodilation in the placenta (Read et al., 1993). One report demonstrates release of angiotensin II by *in vitro* cultured cytotrophoblast cells (Cervar et al., 1994). Of the three endothelins, only ET-1 appears to be made by trophoblast and ET-1 is found predominantly in syncytiotrophoblast, with lesser amounts in vascular endothelium, decidua, and extravillous cytotrophoblast (Malassine et al., 1993). ET-1 binds to both ET_A and ET_B endothelin receptors, which are localized by autoradiography not only to the placental vasculature but also to decidual cells and some extravillous

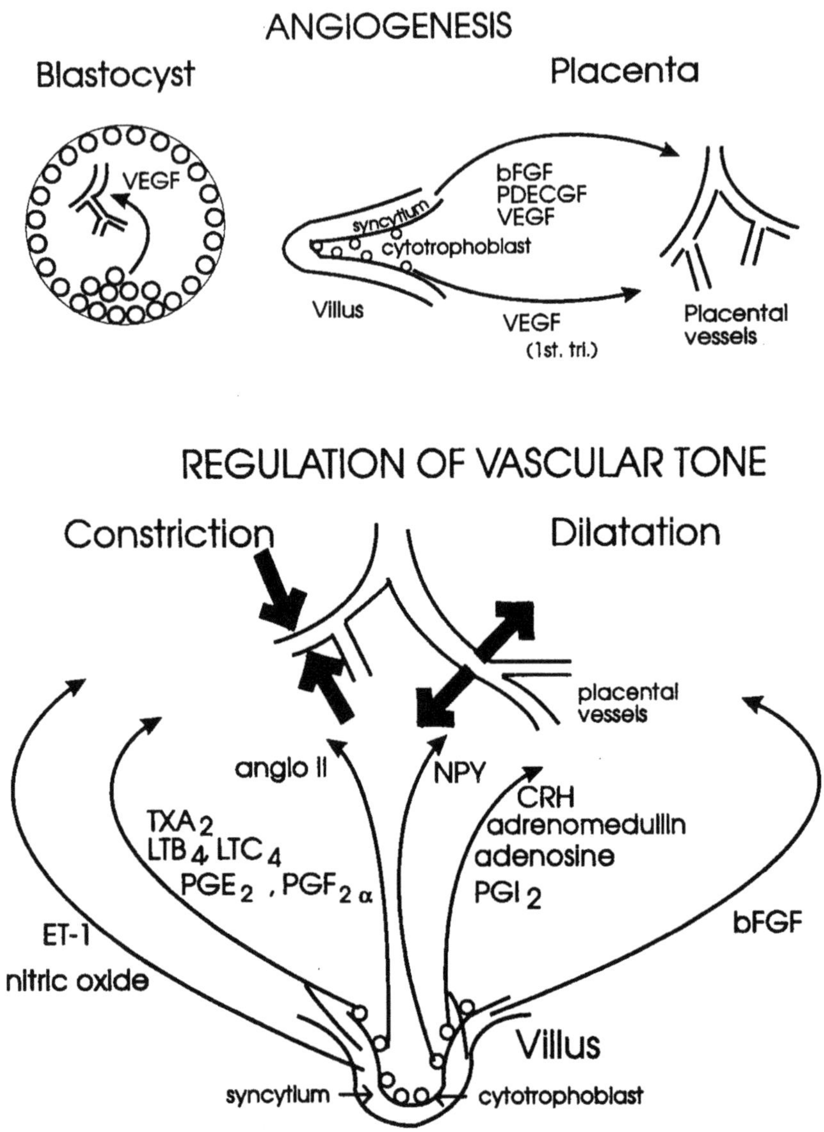

Figure 6. The trophoblast secretes soluble factors involved in regulation of angiogenesis (bFGF, VEGF, PDECGF; upper panel, see the text for the full names). VEGF is secreted by trophectoderm in pre-implantation blastocyst and may induce early vascular formation. Regulation of vascular tone (lower panel) includes both constrictor substances (left side of figure) and dilatory substances (right side).

cytotrophoblast cells (Rutherford et al., 1993), suggesting additional non-vasoactive paracrine functions. Another potent vasodilator is nitric oxide. The calcium-dependent isoform of nitric oxide synthase is present only in endothelium and syncytium in the terminal villi (Myatt et al., 1993a, 1993b; Buttery et al., 1994). Nitric oxide infused into placental lobules causes significant vasodilation (Myatt et al., 1991). It also appears to contribute to maintenance of basal vascular tone and to attenuate the actions of vasoconstrictors in this circulation (Myatt et al., 1992). A recently-recognized vasodilator is corticotropin releasing hormone (CRH). CRH has been localized to both syncytium and cytotrophoblast (Petraglia et al., 1987; Riley et al., 1991). Infusion into placental lobules produces vasodilation 53 times more potent than PGI_2 (Clifton et al., 1994).

Eicosanoid production by the placenta has been intensely studied due to its profound effect on vascular tone and uterine contractility. Eicosanoids are produced from arachidonic acid by three main pathways: the cyclooxygenase pathway producing prostaglandins, prostacyclin, and thromboxanes; the lipoxygenase pathway producing HPETEs, HETEs, leukotrienes (LTs), and lipoxins; and the epoxygenase pathway producing lipoxides (Edwin and Mitchell, 1993). The perfused intact placenta produces massive amounts of these substances, but it has been difficult to determine exactly which cell types in the placenta are the sources (Myatt, 1990; Olson et al., 1993). Cyclooxygenase (prostaglandin endoperoxidase H synthase, PGHS) is present in villous and chorionic cytotrophoblast and villus syncytiotrophoblast (Olson et al., 1993). Placental explants produce 6-keto-$PGF_{2\alpha}$, TXB_2, PGD_2, PGE_2, 15-keto-PGE_2, and other PGE_2 metabolites (Harper et al., 1983; Siler-Khodr and Forman, 1993). Collagenase-dispersed trophoblast cells in monolayer culture produced PGE_1, PGF_1, PGF metabolites, and 6-keto-$PGF_{1\alpha}$ (Olson et al., 1983). Dispase-dispersed trophoblast cells produced mainly diHETEs, LTB_4, monoHETEs, and lesser amounts of PGE_2, PGD_2, epoxide, and thromboxane B_2 (Rose et al., 1987). Collagenase-DNAse dispersed cells produce $PGF_{2\alpha}$ (Pasmanik et al., 1991). However, in none of these three studies was the percentage of trophoblast cells characterized by recognized immunological markers (as used by Yui et al., 1994). Highly purified cytotrophoblast cells in a characterized trophoblast culture system produce thromboxane B_2 and metabolites of thromboxane A_2, LTB_4, and PGI_2 metabolites (Cervar et al., 1994). Overall, there is a predominance of lipoxygenase over cyclooxygenase metabolites. Therefore, it appears likely that the trophoblast is a major source of eicosanoids. Interestingly, the most abundant metabolites have immune and inflammation-related activities rather than vascular effects. However, given the huge placental mass, even the production of a small quantity per cell of a particular eicosanoid implies potentially very significant vascular regulatory effects.

Neuropeptide Y has been localized to the cytotrophoblast (Petraglia et al., 1989) and can cause vasoconstriction. No other studies have been performed with this compound. A recent discovery is the cloning of adrenomedullin, a chromaffin-derived peptide with potent vasodilatory effects (Kitamura et al., 1993a, 1993b). We have recently cloned the human homologue from a placental library and shown that it is present in cytotrophoblast, increasing as syncytial formation occurs (Morrish et al., unpublished data; Figure 3). Future research on this new peptide will elucidate its place among the many other autacoids regulating blood flow.

C. Pathophysiology

The prime vascular disorder of pregnancy is pregnancy-induced hypertension or preeclampsia. Abnormal autacoid production has been implicated in this disease. There is evidence for reduced placental prostacyclin synthesis and increased HETE synthesis, leading to hypertension (reviewed in Myatt, 1990; Edwin and Mitchell, 1993). Abnormal arachidonic acid metabolism may also be implicated in intrauterine growth retardation (Myatt, 1990). Maternal ET-1 and CRH levels are increased in preeclampsia (Laaitkainen et al., 1991; Bobik et al., 1990; McMahon et al., 1993). Nitric oxide synthase activity is reduced in preeclampsia (Brennecke et al., 1994). No studies have been conducted yet on the other vasoactive substances mentioned, but abnormalities in preeclampsia may well be found. In all these studies, it has yet to be proven if the abnormality is primary or secondary.

VI. PARTURITION

This topic has been reviewed in this volume and elsewhere (Myatt, 1990; Keirse, 1990; Casey and MacDonald, 1993; Olson et al., 1993, 1995; Edwin and Mitchell, 1993). In this chapter, we will focus on the question, what is the potential role of the trophoblast in this process? The answer lies in specifying which trophoblast-produced products stimulate myometrial contractility and labor. Trophoblast products perhaps of lesser significance appear to be estrogen or progesterone (Olson et al., 1995), although they may have an indirect effect. Oxytocin, a possible regulator, is not made in the human trophoblast (Chibbar et al., 1993). However, potential factors include eicosanoids, CRH, endothelin, and cytokines (Figure 7).

The eicosanoids are considered the principal final pathway for stimulating endometrial contractility. Current theories favor the decidua or amnion as the major originators of these signals, largely because of the apparent lesser production of cyclooxygenase products from the trophoblast compared to the decidua or amnion (Myatt, 1990; Olson et al., 1995). However, as noted above, the trophoblast as a whole produces large amounts of prostaglandins. Cyclooxygenase products, particularly PGE_2 and $PGF_{2\alpha}$, are the main stimulators of uterine contractility, whereas PGI_2 is a relaxant. As noted above, trophoblast probably makes these compounds. Thus, there is evidence for significant contribution to regulation of uterine contractility from trophoblast-derived prostaglandins.

Considerable work has been directed at trying to demonstrate a role for corticosteroids in parturition (Margioris, 1993). Because administration of even large doses of dexamethasone to pregnant women does not trigger labor, attention has focussed on other actions of the CRH-POMC (pro-opimelanocortin) system in humans. As noted earlier, CRH is made in the trophoblast. CRH may stimulate the fetal adrenal to make cortisol, which induces 17α-hydroxylase, thus promoting estrogen synthesis. The estrogen promotes myometrial gap junction formation (Garfield et al., 1979) and oxytocin receptor expression (Soloff, 1975), and oxytocin gene expression in the chorio-decidua (Chibbar et al., 1995). CRH induces PGE_2 and $PGF_{2\alpha}$ synthesis (Jones and Challis, 1990a, 1990b) and CRH may sensitize the myometrium to the contractile effects of oxytocin (Quartero and Fry, 1989). Prostaglandins may participate in a positive feedback loop by increasing CRH secretion (Petraglia et al., 1987). CRH will also increase syncytiotrophoblast and maternal ACTH secretion, leading to the mild hypercortisolism found late in pregnancy, which may contribute to estrogen synthesis.

Endothelin (ET-1) has complex actions which may have a role in parturition. It induces myometrial, but inhibits amnion and PGE_2 production (Mitchell, 1991) and can induce rhythmic contractions in rat myometrium (Kozuka et al., 1989; King, 1993). It also potently increases myosin light chain phosphorylation in myometrial smooth muscle cells (Word et al., 1990).

Cytokines are likely important in parturition by influencing other processes such as prostaglandin production. For example, chorion, amnion, and decidual PGE_2 production are induced by the trophoblast-produced cytokines IL-1 and TNFα (reviewed in Mitchell et al., 1990, 1993; Chen et al., 1991; Romero et al., 1989). Interestingly, in mice, PGE_2 inhibits production of TH_1 but not TH_2 cytokines (Betz and Fox, 1991). These data are consistent with the theory that TH_2 cytokines favor a successful pregnancy (Wegmann et al., 1993).

In summary, current data support the view that a variety of hormonal influences interact to induce labor through the final common mediator of the prostaglandins. The trophoblast is a major producer of many of these substances and thus must play a significant role in parturition.

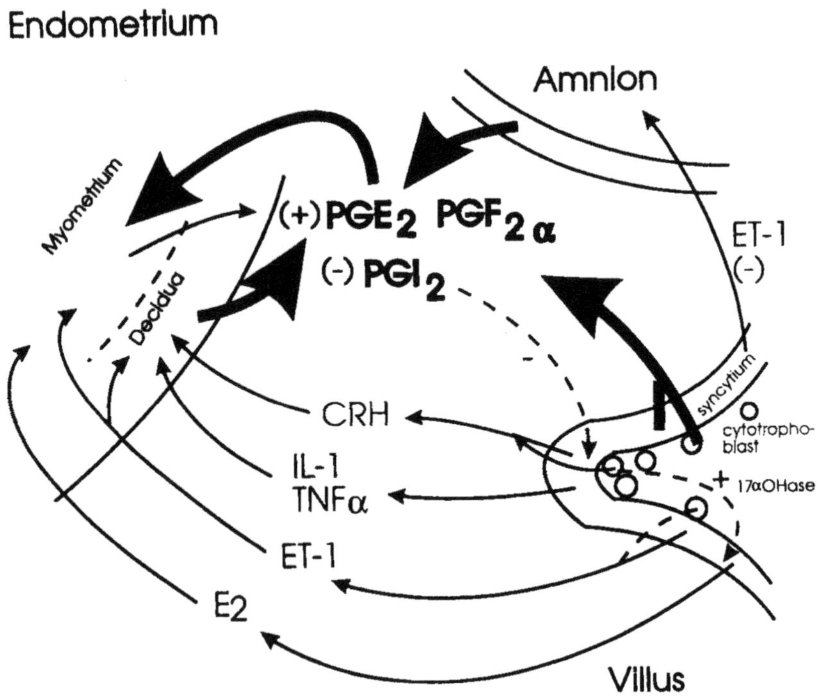

Figure 7. Parturition is a highly complex and regulated process. This figure shows only those substances made by trophoblast that may interact to regulate uterine contractility.

A. Pathophysiology

The cause of preterm labor is a poorly-understood entity. Clearly, any of the factors discussed could induce early labor. One current hypothesis potentially accounting for up to 30% of preterm labor is that infection induces IL-1 and TNFα, which in turn increase prostaglandin production (Radetsky, 1994; Gibbs et al., 1992). There are no reports of a primary abnormality in any of the factors discussed that can be shown to cause preterm labor.

There is one report of elevated CRH in preeclampsia (Laatikainen et al., 1991), but it is unclear how this can be reconciled with a vasodilator effect of CRH. Again, these data describe only an association.

VI. CONCLUSION

In this review, we have asked the question, what evidence is there to support a dominant role of the trophoblast in the physiology and diseases of pregnancy? It is evident that implantation, metabolic regulation, and immune modulation are dominated by the trophoblast. There is accumulating evidence for the trophoblast having a stronger role in angiogenesis and vasomotor tone. Although current theories of parturition favor non-trophoblastic tissues, this attitude may change given the increasingly-recognized role of cytokines, and a careful reconsideration of trophoblast-produced factors. Interestingly, there has yet to be found a definitive primary abnormality of a trophoblast-produced factor or function that results in any disease. As our understanding of normal physiology broadens, it seems likely that such disease relations will be described, considering the lack of successful theories and treatments existing at this time.

REFERENCES

Altman, D.J., Schneider, S.L., Thompson, D.A., Cheng, H.-L., & Tomasi, T.B. (1990). A transforming growth factor β2 (TGF-β2)-like immunosuppressive factor in amniotic fluid and localization of TGFβ2 MRNA in the pregnant uterus. J. Exp. Med. 172, 1391–1401.

Aschheim, S., & Zondek, B. (1927). Anterior pituitary hormone and ovarian hormone in the urine of pregnant women [*Hypophysenvorderlappenhormon und ovarial hormon im Harn von Schwangeren*]. Klin. Wochenschr. 6, 1322.

Autio-Harmainen, H., Hurskainen, T., Niskasaari, K., Hoyhtya, M., & Tryggvason, K. (1992). Simultaneous expression of 70 kilodalton type IV collagenase and type

IV collagen α1 (IV) chain genes by cells of early human placenta and gestational endometrium. Lab. Invest. 67, 191–200.

Autio-Harmainen, H., Sandberg, M., Pihlajaniemi, T., & Vuorio, E. (1991). Synthesis of laminin and type IV collagen by trophoblastic cells and fibroblastic stromal cells in the early human placenta. Lab. Invest. 64, 483–491.

Baird, A. (1993). Editorial: Fibroblast growth factors: What's in a name? Endocrinology 132, 487–488.

Behrendtsen, O., Alexander, C.M., & Werb, Z. (1992). Metalloproteinases mediate extracellular matrix degradation by cells from mouse blastocyst outgrowths. Development 114, 447–456.

Betz, M., & Fox, B.S. (1991). Prostaglandin E_2 inhibits production of Th1 lymphokines but not of Th2 lymphokines. J. Immunol. 146, 108–113.

Bobik, A., Grooms, A., Millar, J.A., Mitchell, A., & Grinpukel, S. (1990). Growth factor activity of endothelin on vascular smooth muscle. Am. J. Physiol. 258, C408–C415.

Bonin, P.D., Leadley, R.J., & Erickson, L.A. (1993). Growth factor-induced modulation of endothelin-1 binding to human smooth-muscle cells. J. Cardiovasc. Pharmacol. 23 (Suppl. 8), S125–S127.

Boura, A.L.A., & Walters, W.A.W. (1991). Autacoids and the control of vascular tone in the human umbilical-placental circulation. Placenta 12, 453–477.

Brennecke, S.P., Dilulio, J.L., Gude, N.M., & King, R.G. (1994). Nitric oxide synthase activity of preterm human placental tissue is reduced in pre-eclampsia. Placenta 15, A6.

Brenner, C.A., Adler, R.R., Rappolee, D.A., Pedersen, R.A., & Werb, Z. (1989). Genes for extracellular matrix-degrading metalloproteinases and their inhibitor, TIMP, are expressed during early mammalian development. Genes Develop. 3, 848–859.

Bulmer, J.N., Morrison, L., Johnson, P.M., & Meager, A. (1990). Immunohistochemical localization of interferons in human placental tissues in normal, ectopic, and molar pregnancy. J. Reprod. Immunol. 22, 109–116.

Buttery, L.D.K., McCarthy, A., Springall, D.R., Sullivan, M.H.F., Elder, M.G., Michel, T., & Polak, J.M. (1994). Endothelial nitric oxide synthase in the human placenta: Regional distribution and proposed regulatory role at the feto-maternal interface. Placenta 15, 257–265.

Cai, W.Q., Terenghi, G., Bodin, P., Burnstock, G., & Polak, J.M. (1993). *In situ* hybridization of atrial natriuretic peptide mRNA in the endothelial cells of human umbilical vessels. Histochemistry 100, 277–283.

Casey, M.L., & MacDonald, P.C. (1993). Cytokines in the human placenta, fetal membranes, uterine decidua, and amniotic fluid. In: *Molecular Aspects of Placental and Fetal Autacoids* (Rice, G.E., & Brennecke, S.P., Eds.) pp. 361–394. CRC Press, Boca Raton, FL.

Cattini, P.A., Nickel, B., Bock, M., & Kardami, E. (1991). Immunolocalization of basic fibroblast growth factor (bFGF) in growing and growth-inhibited placental cells: A possible role for bFGF in placental cell development. Placenta 12, 341–352.

Cedard, L., Alsat, E., Urtasun, M-J., & Varangot, J. (1970). Studies on the mode of action of luteinizing hormone and chorionic gonadotropin on estrogenic biosynthesis and glycogenolysis by human placenta perfused in vitro. Steroids 16, 361–375.

Cervar, M., Kainer, F., & Desoye, G. (1994). Acetylsalicylic acid, but not Mg^{2+} or hydralazine, affects in vitro the release of thromboxane B_2 and 6-keto-prostaglandin-$F_{1\alpha}$ from cultured human trophoblast cells. Placenta 15, A.9.
Chen, E.Y., Liao, Y.-C., Smith, D.H., Barrera-Saldana, H.A., Gelinas, R.E., & Seeburg, P.H. (1989). The human growth hormone locus: Nucleotide sequence, biology, and evolution. Genomics 4, 479–497.
Chen, H.-L., Yang Y., Hu, X.-L., Yelavarthi, K.K., Fishback, J.L., & Hunt, J.S. (1991). Tumor necrosis factor alpha mRNA and protein are present in human placental and uterine cells at early and late stages of gestation. Am. J. Reprod. Immunol. 139, 327–335.
Chibbar, R., Miller, F.D., & Mitchell, B.F. (1993). Synthesis of oxytocin in amnion, chorion, and decidua may influence the timing of human parturition. J. Clin. Invest. 91, 185–192.
Chibbar, R., Wong, S., Miller, F.D., & Mitchell, B.F. (1995). Estrogen stimulates oxytocin gene expression in human chorio-decidua. J. Clin. Endocrinol. Metab. 80, 567–572.
Clark, D.A., Flanders, K.C., Banwatt, D., Millar-Book, W., Manuel, J., Stedronska-Clark, J., & Rowley, B. (1990). Murine pregnancy decidua produces a unique immunosuppressive molecule related to transforming growth factor β-2. J. Immunol. 144, 3008–3014.
Clifton, V.L., Read, M.A., Leitch, I.M., Boura, A.L.A., Robinson, P.J., & Smith, R. (1994). Corticotropin-releasing hormone-induced vasodilatation in the human fetal placental circulation. J.Clin. Endocrinol. Metab. 79, 666–669.
Cross, J.C., Werb, Z., & Fisher, S.J. (1994). Implantation and the placenta: Key pieces of the development puzzle. Science 266, 1508–1518.
Cuevas, P., Carceller, F., Ortega, S., Zazo, M., Nieto, I., & Gimenez-Gallego, G. (1991). Hypotensive activity of fibroblast growth factor. Science 254, 1208–1210.
Daiter, E., Pampfer, S., Yeung, Y.G., Barad, D., Stanley, E.R., & Pollard, J.W. (1992). Expression of colony-stimulating factor-1 in the human uterus and placenta. J. Clin. Endocrinol. Metab. 74, 850–858.
Damsky C.H., Fitzgerald M.L., & Fisher S.J. (1992). Distribution patterns of extracellular matrix components and adhesion receptors are intricately modulated during first trimester cytotrophoblast differentiation along the invasive pathway, *in vivo*. J. Clin. Invest. 89, 210–222.
Edwin, S.S., & Mitchell, M.D. (1993). Arachidonic acid metabolism via lipoxygenase pathways in the placenta and fetal membranes. In: *Molecular Aspects of Placental and Fetal Membrane Autacoids* (Rice, G.E., & Brennecke, S.P., Eds.) pp. 127–156. CRC Press, Boca Raton, FL.
Ellis, S. (1990). HLA-G: At the interface. Am. J. Reprod. Immunol. 23, 84–86.
Emonard, H., Christiane, Y., Smet, M., Grimaud, J.A., & Foidart, J.M. (1990). Type IV and interstitial collagenolytic activities in normal and malignant trophoblast cells are specifically regulated by the extracellular matrix. Invasion Metast. 10, 170–177.
Estelles, A., Keeton, J.G.M., Eguchi, Y., Aznar, J., Grancha, S., Espana, F., Loskutoff, D.J., & Schleef, R.R. (1994). Altered expression of plasminogen activator inhibitor type 1 in placentas from pregnant women with preeclampsia and/or intrauterine fetal growth retardation. Blood 84, 143–150.

Falcone, T., & Little, A.B. (1994). Maintenance of pregnancy and the initiation of normal labor. In: *Maternal-Fetal Endocrinology* (Tulchinsky, D., & Little, A.B., Eds.) pp. 351–363. W.B. Saunders Co., Philadelphia.

Feinberg, R.F., Kao, L.-C., Haimowitz J.E., Queenan, J.T., Wun, T.-C., Strauss, J.F., & Kliman, H.J. (1989). Plasminogen activator inhibitor types 1 and 2 in human trophoblasts. Lab. Invest. 61, 20–26.

Feinberg, R.F., Kliman, H.J., & Lockwood, C.J. (1991). Is oncofetal fibronectin a trophoblast glue for human implantation? Am. J. Pathol. 138, 537–543.

Fernandez, P.L., Merino, M.J., Nogales, F.F., Charonis, A.S., Stetler-Stevenson, W. & Liotta, L. (1992). Immunohistochemical profile of basement membrane proteins and 72 kilodalton type IV collagenase in the implantation placental site. Lab. Invest. 66, 572–579.

Ferry, B.L., Sargent, I.L., Starkey, P.M., & Redman, C.W.G. (1991). Cytotoxic activity against trophoblast and choriocarcinoma cells of large granular lymphocytes from human early pregnancy decidua. Cell. Immunol. 132, 140–149.

Fisher, S.J., Leitch, M.S., Kantor, M.S., Basbaum, C.B., & Kramer, R.H. (1985). Degradation of extracellular matrix by the trophoblastic cells of first-trimester human placentas. J. Cell. Biochem. 27, 31–41.

Fisher, S.J., Cui, T.-Y., Zhang, L., Hartman, L., Grahl, K., Zhang, G.-Y., Tarpey, J., & Damsky, C.H. (1989). Adhesive and degradative properties of human placental cytotrophoblast cells *in vitro*. J. Cell Biol. 109, 891–902.

Flaumenhaft, R., Abe, M., Mignatti, P., & Rifkin, D.B. (1992). Basic fibroblast growth factor-induced activation of latent transforming growth factor β in endothelial cells: Regulation of plasminogen activator activity. J. Cell Biol. 118, 901–909.

Folkman, J. (1995). Angiogenesis in cancer, vascular, rheumatoid and other disease. Nature Med. 1, 27–31.

Frankenne, F., Rentier-Delrue, F., Scippo, M.-L., Martial, J., & Hennen, G. (1987). Expression of the growth hormone variant gene in human placenta. J. Clin. Endocrinol. Metab. 64, 635–637.

Garcia-Lloret, M., Morrish, D.W., Wegmann, T.G., Honore, L., Turner, A.R., & Guilbert, L.J. (1994). Demonstration of functional cytokine-placental interactions: CSF-1 and GM-CSF stimulate human cytotrophoblast differentiation and peptide hormone secretion. Exp. Cell Res. 214, 46–54.

Garfield, R.E., Rabideau, S., Challis, J.R.G., & Daniel, E.E. (1979). Hormonal control of gap junctions in sheep myometrium. Biol. Reprod. 21, 999–1007.

Gibbs, R.S., Romero, R., Hillier, S.L., Eschenbach, D.A., & Sweet, R.L. (1992). A review of premature birth and subclinical infection. Am. J. Obstet. Gynecol. 166, 1515–1528.

Glasser, S.R., Julian, J.A., Decker, G.L., Tang, J-Y., & Carson, D.D. (1988). Development of morphological and functional polarity in primary cultures of immature rat uterine epithelial cells. J. Cell Biol. 107, 2409–2423.

Goldyne, M.E., & Stobo, J.D. (1981). Immunoregulatory role of prostaglandins and related lipids. Crit. Rev. Immunol. 2, 189–224.

Goodfellow, C.F., & Taylor, P.V. (1982). Extraction and identification of trophoblast cells circulating in peripheral blood during pregnancy. Br. J. Obstet. Gynecol. 89, 65–68.

Gospodarowicz, D., Cheng, J., Lui, G.-M., Fujii, D.K., Baird, A., & Bohlen, P. (1985). Fibroblast growth factor in the human placenta. Biochem. Biophys. Res. Commun. 128, 554–562.

Gossrau, R., Graf, R., Ruhnke, M., & Hanski, C. (1987). Proteases in the human full-term placenta. Histochemistry 86, 405–413.

Graham, C.H., & Lala, P.K. (1992). Mechanisms of placental invasion of the uterus and their control. Biochem. Cell Biol. 70, 867–874.

Grainger, D.J., Kemp, P.R., Metcalf, J.C., Liu, A.C., Lawn, R.M., Williams, N.R., Grace, A.A., Schofield, P.M., & Chauhan, A. (1995). The serum concentration of active transforming growth factor-β is severely depressed in advanced atherosclerosis. Nature Med. 1, 74–79.

Gris, J.-C., Neveu, S., Mares, P., Biron, C., Hedon, B., & Schved, J.-F. (1993). Plasma fibrinolytic activators and their inhibitors in women suffering from early recurrent abortion of unknown etiology. J. Lab. Clin. Med. 122, 606–615.

Halban, J. (1905). The internal secretions of the ovaries and placenta and their importance for the function of the mammary glands in pregnancy [*Die innere Secretion von Ovarium und Placenta und ihre Bedeutung fur die Function der Milchdruse*]. Archiv fur Gynakologie. 75, 353–411.

Harper, M.J.K., Khodr, G.S., & Valenzuela, G. (1983). Prostaglandin production by human term placentas in vitro. Prostaglandins Leukotrienes and Medicine 11, 121–129.

Harris, S.J., Anthony, F.W., Jones, D.B., & Masson, G.M. (1984). Pregnancy-specific-β1-glycoprotein: Effect on lymphocyte proliferation *in vitro*. J. Reprod. Immunol. 6, 267–270.

Hennen, G., Frankenne, F., Closset, J., Gomez, F., Pirens, G., & El Khayat, N. (1985). A human placental GH: Increasing levels during second half of pregnancy with pituitary GH suppression as revealed by monoclonal antibody radioimmunoassays. Int. J. Fertil. 30, 27–33.

Hertig, A.T., & Rock, J. (1945). Two human ova of the pre-villous stage, having a developmental age of about seven and nine days respectively. Contrib. Embryol. 31, 67–86.

Herz, J., Clouthier, D.E., & Hammer, RE. (1992). LDL receptor-related protein internalizes and degrades uPA-PAI-1 complexes and is essential for embryo implantation. Cell 71, 411–421.

Hill, D.J. (1992). What is the role of growth hormone and related peptides in implantation and the development of the embryo and fetus. Horm. Res. 38 (Suppl. 1), 28–34.

Hill, J.A. (1994). Sporadic and recurrent spontaneous abortion. Curr. Probl. Obstet. Gynecol. Fertil. 17, 113–164.

Hofmann, G.E., Glatstein, I., Schatz, F., Heller, D., & Deligdisch, L. (1994). Immunohistochemical localization of urokinase-type plasminogen activator and the plasminogen activator inhibitors 1 and 2 in early human implantation sites. Am. J. Obstet. Gynecol. 170, 671–676.

Hoshina, M., Boothby, M., & Boime, I. (1982). Cytological localization of chorionic gonadotropin α and placental lactogen mRNA during development of the human placenta. J. Cell Biol. 93, 190–198.

Hoshina, M., Hussa, R., Pattillo, R., & Boime, I. (1983). Cytological distribution of chorionic gonadotropin subunit and placental lactogen messenger RNA in neoplasms derived from human placenta. J. Cell Biol. 97, 1200–1206.

Huhtaniemi, I.T., Korenbrot, C.C., & Jaffee, R.B. (1977). hCG binding and stimulation of testosterone biosynthesis in the human fetal testis. J. Clin. Endocrinol. Metab. 44, 963–967.

Hunt, J.S., King, C.R., & Wood, G.W. (1984). Evaluation of human chorionic trophoblast cells and placental macrophages as stimulators of maternal lymphocyte proliferation in vitro. J. Reprod. Immunol. 6, 377–391.

Hunt, J.S., & Hsi, B-L. (1990). Evasive strategies of trophoblast cells: Selective expression of membrane antigens. Am. J. Reprod. Immunol. 23, 57–63.

Hynes, R.O. (1992). Integrins: Versatility, modulation, and signaling in cell adhesion. Cell 69, 11–25.

Irving, J.A., & Lala, P.K. (1994). Decidua-derived IGFBP-1 stimulates human intermediate trophoblast migration by binding to the $\alpha5/\beta1$ integrin subunits. Placenta 15, A.31.

Jaameri, K.E.U., Koivuniemi, A., Carpen, E.O. (1965). Occurrence of trophoblasts in the blood of toxemic patients. Gynaecologica 160, 315–320.

Jackson, M.R., Carney, E.W., Lye, S.J., & Knox Ritchie, J.W. (1994). Localization of two angiogenic growth factors (PDECGF and VEGF) in human placentae throughout gestation. Placenta 15, 341–353.

Jakeman, L.B., Armanini, M., Phillips, H.S., & Ferrara, N. (1993). Developmental expression of binding sites and messenger ribonucleic acid for vascular endothelial growth factor suggests a role for this protein in vasculogenesis and angiogenesis. Endocrinology, 133, 848–859.

Jones, C.T. (1993). Endocrine and metabolic interaction between placenta and fetus: Pathways of maternal-fetal communication. In: *The Human Placenta* (Redman, C.W.G., Sargent I.L., & Starkey, F.M., Eds.) pp. 527–557. Blackwell Scientific Publications, London.

Jones, S.A., & Challis, J.R.G. (1990a). Steroid, corticotrophin-releasing hormone, ACTH, and prostaglandin interaction in the amnion and placenta of early pregnancy in man. J.Endocrinol. 125, 153–159.

Jones, S.A., & Challis, J.R.G.(1990b). Effects of corticotropin-releasing hormone and adrenocorticotropin on prostaglandin output by human placenta and fetal membranes. Gynecol. Obstet. Invest. 29, 165–168.

Keirse, M.J.N.C. (1990). Eicosanoids in human pregnancy and parturition. In: *Eicosanoids in Reproduction* (Mitchell, M.D., Ed.) pp. 199–222. CRC Press, Boca Raton, FL.

Kelly, A.M., England P., Lorimer, J.D., & Ferguson, J.C. (1975). An evaluation of human placental lactogen levels in hypertension of pregnancy. Br. J. Obstet. Gynecol. 82, 272–277.

Khalfoun, B., Degenne, D., Crouzat-Reynes, G., & Bardes, P. (1986). Effect of human syncytiotrophoblast plasma membrane-soluble extracts on in vitro mitogen-induced lymphocyte proliferation. A possible inhibition mechanism involving the transferrin receptor. J. Immunol. 137, 1187–1193.

Kido, I. (1937).] The human placenta is a manufacturing plant of the so-called anterior lobe pituitary hormones (experimental investigations). [*Die menschliche Placenta als produktionsstatte des sogenannten Hypophysonvorderlappenhormons (experimentelle Untersuchungen)*]. Zentralblattfur Gynakologie 61, 1551–1555.

King, R.G. (1993). Endothelins and the placenta. In: *Molecular Aspects of Placental and Fetal Membrane Autacoids* (Rice, G.E., & Brennecke, S.P., Eds.) pp. 251–261. CRC Press, Boca Raton, FL.

Kitamura, K., Kanagawa, K., Kawamoto, M., Ichiki, Y., Nakamura, S., Matsuo, H., & Eto, T. (1993a). Adrenomedullin: A novel hypotensive peptide isolated from human pheochromocytoma. Biochem. Biophys. Res. Commun. 192, 553–560.

Kitamura, K., Sakata J., Kanagawa, K., Kojima, M., Matsuo, H., & Eto, T. (1993b). Cloning and characterization of cDNA encoding a precursor for human adrenomedullin. Biochem. Biophys. Res. Commun. 194, 720–725.

Korhonen, M., Ylanne, J., Laitinen, L., Cooper, H.M., Quaranta, V., & Virtanen, I. (1991). Distribution of the α1–α6 integrin subunits in human developing and term placenta. Lab. Invest. 65, 347–356.

Kovats, S., Main, E.K., Librach, C., Stubblebine, M., Fisher, S.J., & DeMars, R. (1990). A class I antigen, HLA-G, expressed in human trophoblasts. Science 248, 220–223.

Kozuka, M., Ito, T., Hirose, S., Takahashi, K., & Hagiwara, H. (1989). Endothelin induces two types of contractions of rat uterus: Phasic contractions by way of voltage-dependent calcium channels and developing contractions through a second type of calcium channels. Biochem. Biophys. Res. Commun. 159, 317–323.

Kurman, R.J., Main, C.S., & Chen, H.-C. (1984). Intermediate trophoblast: A distinctive form of trophoblast with specific morphological, biochemical and functional features. Placenta 5, 349–369.

Laatikainen, T., Virtanen, T., Kaaja, R., & Salminen-Lappalainen, K. (1991). Corticotropin-releasing hormone in maternal and cord plasma in pre-eclampsia. Eur. J. Obstet. Gynecol. Reprod. Biol. 39, 19–24.

Lala, P.K., & Graham, C.H. (1990). Mechanisms of trophoblast invasiveness and their control: The role of proteases and protease inhibitors. Cancer Metastasis Rev. 9, 369–379.

Le, S.Q., Wasserstrum, N., Mombouli, J.V., & Vanhoutte, P.M. (1993). Contractile effect of endothelin in human placental veins: Role of endothelium prostaglandins and thromboxane. Am. J. Obstet. Gynecol. 169, 919–924.

Lea, R.G., & Clark, D.A. (1991). Macrophages and migratory cells in endometrium relevant to implantation. Baill. Clin. Obstet. Gynecol. 5, 25–59.

Lea, R.G., Harley, C.B., & Clark, D.A. (1990). The detection of RNA for TGFβ2 in post-implantation decidual tissue correlates with the detection of a pregnancy associated suppressor. Eur. Fed. Immunol. Soc., 10th meeting, British Society for Immunology, Edinburgh.

Lea, R.G., Flanders, K.C., Harley, C.B., Manuel, J., Banwatt, D., & Clark, D.A. (1992). Release of a transforming growth factor (TGF)-β2-related suppressor factor from postimplantation murine decidual tissue can be correlated with the detection of a subpopulation of cells containing RNA for TGFβ2. J. Immunol. 148, 778–787.

Lessey, B.A., Damjanovich, L., Coutifaris, C., Castelbaum, A., Albelda, S.M., & Buck, C.A. (1992). Integrin adhesion molecules in the human endometrium. J. Clin. Invest. 90, 188–195.

Librach, C.L., Werb, Z., Fitzgerald, M.L., Chiu, K., Corwin, N.M., Esteves, R.A., Grobelny, D., Galardy, R., Damsky, C.H., & Fisher, S.J. (1991). 92-kD type IV collagenase mediates invasion of human cytotrophoblasts. J. Cell Biol. 113, 437–449.

Lindhoff, C., & Astedt, B. (1994). Plasminogen activator of urokinase type and its inhibitor of placental type in hypertensive pregnancies and in intrauterine growth retardation: Possible markers of placental function. Am. J. Obstet. Gynecol. 171, 60–64.

Liotta, L.A. (1990). Introductory overview. Cancer Metastasis Rev. 9, 285–287.

Liotta, L. (1992). Immunohistochemical profile of basement membrane proteins and 72 kilodalton type IV collagenase in the implantation site. Lab. Invest. 66, 572–579.

Lyons, R.M., Keski-Oja, J., & Moses, H.L. (1988). Proteolytic activation of latent transforming growth factor-β from fibroblast conditioned medium. J. Cell Biol. 106, 1659–1665.

Malassine, A., Cronier, L., Mondon, F., Mignot, T.M., & Ferre, F. (1993). Localization and production of immunoreactive endothelin-1 in the trophoblast of human placenta. Cell Tissue Res. 271, 491–497.

Margioris, A.N. (1993). Corticotropin-releasing hormone and propiomelanocortin in placenta and fetal membranes. In: *Molecular Aspects of Placental and Fetal Membrane Autacoids* (Rice, G.E. & Brennecke, S.P., Eds.) pp. 277–301. CRC Press, Boca Raton, FL.

Martin, O., & Arias, F. (1982). Plasminogen activator production by trophoblast cells *in vitro*: Effect of steroid hormones and protein synthesis inhibitors. Am. J. Obstet. Gynecol. 142, 402–409.

Matsuzaki, N., Li, Y., Masuhiro, K., Jo, T., Shimoya, K., Taniguchi, T., Saji, F., & Tanizawa, O. (1992). Trophoblast-derived transforming growth factor-β1 suppresses cytokine-induced, but not gonadotropin-releasing hormone-induced, release of human chorionic gonadotropin by normal human trophoblasts. J. Clin. Endocrinol. Metab. 74, 211–216.

McMahon, L.P., Redman, C.W.G., & Firth, J.D. (1993). Expression of the three endothelin genes and plasma levels of endothelin in pre-eclamptic and normal gestations. Clinical Science 85, 417–424.

McWilliams, D., & Boime, I. (1980). Cytological localization of placental lactogen messenger ribonucleic acid in syncytiotrophoblast layers of human placenta. Endocrinology 107, 761–765.

Mitchell, M.D., Edwin, S., & Romero, R.J. (1990). Prostaglandin biosynthesis by human decidual cells: Effects of inflammatory mediators. Prostaglandins Leukotrienes EFA. 41, 35–38.

Mitchell, M.D. (1991). Endothelins in perinatal biology. Semin. Perinatol. 15, 79–85.

Mitchell, M.D., Trautman, M.S., & Dudley, D.J. (1993). Cytokine networking in the placenta. Placenta 14, 249–275.

Morrish, D.W., Bhardwaj, D., Dabbagh, L.K., Marusyk, H., & Siy, O. (1987). Epidermal growth factor induces differentiation and secretion of human chorionic gonadotropin and placental lactogen in normal human placenta. J. Clin. Endocrinol. Metab. 65, 1282–1290.

Morrison, J.J., Clark, I.M., Powell, E.K., Cawston, T.E., Hackett, G.A., & Smith, S.K. (1994). Tissue collagenase: Serum levels during pregnancy and parturition. Eur. J. Obstet. Gynecol. 54, 71–75.

Myatt, L. (1990). Placental biosynthesis, metabolism, and transport of eicosanoids. In: *Eicosanoids in Reproduction* (Mitchell, M.D., Ed.) pp. 169–198. CRC Press, Boca Raton, FL.

Myatt, L. (1992). Current topic: Control of vascular resistance in the human placenta. Placenta 13, 329–341.

Myatt, L., Brewer, A., & Brockman, D.E. (1991). The action of nitric oxide in the perfused human fetal-placental circulation. Am. J. Obstet. Gynecol. 164, 687–692.

Myatt, L., Brewer, A., Langdon, G., & Brockman, D.E. (1992). Attenuation of the vasoconstrictor effects of thromboxane and endothelin by nitric oxide in the human fetal-placental circulation. Am. J. Obstet. Gynecol. 166, 224–230.

Myatt, L., Brockman, D.E., Eis, A.L., & Pollack, J.S. (1993a). Immunohistochemical localization of nitric oxide synthase in the human placenta. Placenta 14, 487–495.

Myatt, L., Brockman, D.E., Langdon, G., & Pollack, J.S. (1993b). Constitutive calcium-dependent isoform of nitric oxide synthase in the human placental villous vascular tree. Placenta 14, 373–383.

Oberauer, A.M., Linkhart, T.A., Mohan, S., & Longo, L.D. (1988). Fibroblast growth factor enhances human chorionic gonadotropin synthesis independent of mitogenic stimulation in Jar choriocarcinoma cells. Endocrinology 123, 2696–2700.

Olson, D.M., Skinner, K., & Challis, J.R.G. (1983). Prostaglandin output in relation to parturition by cells dispersed from human intrauterine tissues. J. Clin. Endocrinol. Metab. 57, 694–699.

Olson, D.M., Zakar, T., & Mitchell, B.F. (1993). Prostaglandin synthesis regulation by intrauterine tissues. In: *Molecular Aspects of Placental and Fetal Membrane Autacoids* (Rice, G.E. & Brennecke, S.P., Eds.) pp. 55–96. CRC Press, Boca Raton, FL.

Olson, D.M., Mijovic, J.E., & Sadowsky, D.W. (1995). Control of human parturition. Semin. Perinatol. 19, 52–63.

Osmers, R., Rath, W., Adelmann-Grill, B.C., Fittkow, C., Kuloczik, M., Szeverenyi, M., Tschesche, H., & Kuhn, W. (1992). Origin of cervical collagenase during parturition. Am. J. Obstet. Gynecol. 166, 1455–1460.

Pasmanik, M., Izhar, M., Zolti, M., & Shemesh, M. (1991). Dexamethasone inhibition of prostaglandin production in human term placental cells is protein and ribonucleic acid synthesis dependent. Endocrinology 128, 2058–2064.

Pelliniemi, L.J., & Dym. M. (1980). The fetal gonad and sexual differentiation. In: *Maternal-Fetal Endocrinology* (Tulchinsky, D., & Ryan, K.J., Eds.) pp. 252–280. W.B.Saunders Co., Philadelphia.

Petraglia, F., Sawchenko, P.E., Rivier, J., & Vale, W. (1987). Evidence for local stimulation of ACTH secretion by corticotropin-releasing hormone in human placenta. Nature 328, 717–719.

Petraglia, F., Calza, L., Giardino, L., Sutton, S., Marrama, P., Rivier, J., Genazzani, A.R., & Vale, W. (1989). Identification of immunoreactive neuropeptide-Y in human placenta: Localization, secretion, and binding sites. Endocrinology 124, 2016–2022.

Puistola, U., Ronnberg, L., Martikainen, H., & Turpeenniemi-Hujanen, T. (1989). The human embryo produces basement membrane collagen (type IV collagen)-degrading protease activity. Hum. Reprod. 4, 309–311.

Purcell, D.F.J., McKenzie, I.F.C., Lublin, D.M., Johnson, P.M., Atkinson, J.P., Oglesby, T.J., & Deacon, N.J. (1990). The human cell surface glycoproteins HuLy-m5, membrane cofactor protein (MCP) of the complement system, and trophoblast leucocyte-common (TLX) antigen, are CD46. Immunology 70, 155–161.

Quartero, H.W., & Fry, C.H. (1989). Placental corticotropin-releasing factor may modulate human parturition. Placenta 10, 439–444.

Queenan, J.T., Kao, L.-C., Arboleda, C.E., Ulloa-Aguirre, A., Golos, T.G., Cines, D.B., & Strauss, J.F. (1987). Regulation of urokinase-type plasminogen activator production by cultured human cytotrophoblasts. J. Biol. Chem. 262, 10903–10906.

Radetsky, P. (1994). Stopping premature births before it's too late. Science 266, 1486–1488.

Rajabi, M., Dean, D.D., & Woessner, J.F. (1987). High levels of serum collagenase in premature labor—a potential biochemical marker. Obstet. Gynecol. 69, 179–186.

Rajabi, M.R., Dean, D.D., & Woessner, J.F. (1990). Changes in active and latent collagenase in human placenta around the time of parturition. Am. J. Obstet. Gynecol. 163, 499–505.

Read, M.A., Boura, A.L.A., & Walters, W.A.W. (1993). Vascular actions of purines in the foetal circulation of the human placenta. Br. J. Pharmacol. 110, 454–460.

Reith, A., Moore, N.R., & Bennett, B. (1993). Plasminogen activator inhibitors (PAI-1 and PAI-2) in normal pregnancies, pre-eclampsia and hydatidiform mole. Br. J. Obstet. Gynecol. 100, 370–374.

Riley, S.C., Walton, J.C., Herlick, J.M., & Challis, J.R.G. (1991). The localization and distribution of corticotropin-releasing hormone in the human placenta and fetal membranes throughout gestation. J. Clin. Endocrinol. Metab. 72, 1001–1007.

Romero, R., Durum, S., Dinarello, C.A., Oyarzun, E., Hobbins, J.C., & Mitchell, M.D. (1989). Interleukin-1 stimulates prostaglandin biosynthesis by human amnion. Prostaglandins 37, 13–22.

Rose, M.P., Elder, M.G., & Myatt, L. (1987). Arachidonic acid metabolism in the human placenta. Troph. Res. 2, 71–83.

Rutherford, R.A.D., Wharton, J., McCarthy, A., Gordon, L., Sullivan, M.H.F., Elder, M.G., & Polak, J.M. (1993). Differential localization of endothelin ET_A and ET_B binding sites in human placenta. Br. J. Pharmacol. 109, 544–552.

Saito, S., Kato, Y., Maruyama, M., & Ichijo, M. (1992). A study of interferon-β and interleukin-2 production in premature neonates and neonates with intrauterine growth retardation. Am. J. Reprod. Immunol. 27, 63–68.

Sappino, A.-P., Huarte, J., Belin, D., & Vassalli, J.-D. (1989). Plasminogen activators in tissue remodeling and invasion: mRNA localization in mouse ovaries and implanting embryos. J. Cell Biol. 109, 2471–2479.

Sargent, I.L., & Redman, C.W.G. (1989). Maternal immune responses to the fetus in human pregnancy. In: *Immunology of Pregnancy and Its Disorders* (Stern, C.M.M., Ed.) pp. 115–142. Kluwer, Dordecht.

Sargent, I.L., Redman, C.W.G., & Starkey, P.M. (1994). The placenta as a graft. In: *The Human Placenta* (Redman, C.W.G., Sargent, I.L., & Starkey, P.M., Eds.) pp. 334–361. Blackwell Scientific Publications, Oxford.

Sasagawa, M., Yamazaki, T., Endo, M., Kanazawa, K., & Takeuchi, S. (1987). Immunohistochemical localization of HLA antigens and placental proteins (αhCG, βhCG CTP, hPL and SP1) in villous and extravillous trophoblast in normal human pregnancy: A distinctive pathway of differentiation of extravillous trophoblast. Placenta 8, 515–528.

Seeburg, P.H. (1989). The human growth hormone locus: Nucleotide sequence, biology, and evolution. Genomics 4, 479–497.

Seron-Ferre, M., Lawrence, C.C., & Jaffe, R.B. (1978). Role of hCG in the regulation of the fetal zone of the human fetal adrenal gland. J. Clin. Endocrinol. Metab. 46, 834–837.

Shifren, J.L., Doldi, N., Ferrara, N., Mesiano, S., & Jaffe, R.B. (1994). In the human fetus, vascular endothelial growth factor is expressed in epithelial cells and myocytes, but not vascular endothelium: Implications for mode of action. J. Clin. Endocrinol. Metab. 79, 316–322.

Shimonovitz, S., Hurwitz, A., Dushnik, M., Anteby, E., Geva-Eldar, T., & Yagel, S. (1994). Developmental regulation of the expression of 72 and 92 kd type IV collagenases in human trophoblasts: A possible mechanism for control of trophoblast invasion. Am. J. Obstet. Gynecol. 171, 832–838.

Siler-Khodr, T.M., & Forman, J. (1993). The effect of IGF-I on placental prostanoid production. Prostaglandins 46, 361–370.

Simon, P., Decoster, C., Brocas, H., Schwers, J., & Vassart, G. (1986). Absence of human chorionic somatomammotropin during pregnancy associated with two types of gene deletion. Hum. Genet. 74, 235–238.

Soloff, M.S. (1975). Uterine receptors for oxytocin: Effects of estrogen. Biochem. Biophys. Res. Commun. 65, 205–212.

Spellacy, W.N., Buhti, W.C., & McCreary, S.A. (1974). Measurement of human placental lactogen with a simple immunodiffusion kit (Plac Gest). Obstet. Gynecol. 43, 306–309.

Spivack, M. (1943). On the presence or absence of nerves in the umbilical blood vessels of man and guinea pig. Anatom. Rec. 85, 85–109.

Starkey, P.M. (1994). The decidua and factors controlling implantation. In: *The Human Placenta* (Redman, C.W.G., Sargent, I.L., & Starkey, P.M., Eds.) pp. 362–413. Blackwell Scientific Publications, Oxford.

Starkey, P.M., Sargent, I.L., & Redman, C.W.G. (1988). Cell populations in human early pregnancy decidua: Characterization and isolation of large granular lymphocytes by flow cytometry. Immunology 65, 129–134.

Strickland, S., Reich, E., & Sherman, M.I. (1976). Plasminogen activator in early embryogenesis: Enzyme production by trophoblast and parietal endoderm. Cell 9, 231–240.

Strickland, S., & Richards, W.G. (1992). Invasion of the trophoblasts. Cell. 71, 355–357.

Tabarelli, M., Kofler, R., & Wick, G. (1983). Placental hormones: I. Immunofluorescence studies of the localization of chorionic gonadotrophin, placental lactogen and prolactin in human and rat placenta and in the endometrium of pregnant rats. Placenta 4, 379–388.

Turpeenniemi-Hujanen, T., Ronnberg, L., Kauppila, A., & Puistola, U. (1992). Laminin in the human embryo implantation: Analogy to the invasion by malignant cells. Fertil. Steril. 58, 105–113.

Testa, J.E., & Quigley, J.P. (1991). Reversal of misfortune: TIMP-2 inhibits tumor cell invasion. JNCI 83, 740–742.

Usuki, K., Norberg, L., Larsson, E., Miyazono, K., Hellman, U., Wernstedt, C., Rubin, K., & Heldin, C-H. (1990). Localization of platelet-derived endothelial cell growth factor in human placenta and purification of an alternatively processed form. Cell Regulation. 1, 577–596.

Vaheri, A., Stephens, R.W., Salonen, E.-M., Pollanen, J., & Tapiovaara, H. (1990). Plasminogen activation at the cell surface-matrix interface. Cell Differ. Dev. 32, 255–262.

Walker, D.W., & MacLean, J.R. (1971). Absence of adrenergic nerves in human placenta. Nature 229, 344–345.

Walters, W.A.W. (1992). Autacoids affecting vascular tone in the human fetal extracorporeal circulation. Clin. Exp. Pharmacol. Physiol. 19, 3–8.

Wegmann, T.G., Lin, H., Guilbert, L., & Mosmann, T.R. (1993). Bidirectional cytokine interactions in the maternal-fetal relationship: Is successful pregnancy a TH_2 phenomenon? Immunol. Today 14, 353–356.

Weir, M.R., Li, X.W., Gomolka, D., Peppler, R., O'Bryan-Tear, C.G., & Moran, M.A. (1991). The immunosuppressive properties of enisoprost and a 5-lipoxygenase inhibitor (SC-45662). Transplantation 52, 1053–1057.

Werb, Z., Tremble, P., & Damsky, C.H. (1990). Regulation of extracellular matrix degradation by cell-extracellular matrix interactions. Cell Differ. Dev. 32, 299–306.

Word, R.A., Kamm, K.E., Stull, J.T., & Casey, M.L. (1990). Endothelin increases cytoplasmic calcium and myosin phosphorylation in human myometrium. Am. J. Obstet. Gynecol. 162, 1103–1108.

Yagel, S., Geva, T.E., Solomon, H., Shimonovitz, S., Reich, R., Finci-Yeheskel, Z., Mayer, M., & Milwidsky, A. (1993). High levels of human chorionic gonadotropin retard first trimester trophoblast invasion *in vitro* by decreasing urokinase plasminogen activator and collagenase activities. J. Clin. Endocrinol. Metab. 77, 1506–1511.

Yui, J., Garcia-Lloret, M., Brown, A.J., Berdan, R.C., Morrish, D.W., Wegmann, T.G., & Guilbert, L.J. (1994). Functional, long-term cultures of human term trophoblasts purified by column-elimination of CD9 expressing cells. Placenta 15, 231–246.

Zini, J.-M., Murray, S.C., Graham, C.H., Lala, P.K., Kariko, K., Barnathan, E.S., Mazar, A., Henkin, J., Cines, D.B., & McCrae, K.R. (1992). Characterization of urokinase receptor expression by human placental trophoblasts. Blood 79, 2917–2929.

Zhou, Y., Damsky, C.H., Chiu, K., Roberts, J.M., & Fisher, S.J. (1993). Preeclampsia is associated with abnormal expression of adhesion molecules by invasive cytotrophoblasts. J. Clin. Invest. 91, 950–960.

THE ENDOCRINOLOGY OF LATE PREGNANCY AND PARTURITION

Tamas Zakar and Bryan F. Mitchell

ABSTRACT

The establishment, maintenance, and termination of pregnancy is dependent on endocrine and paracrine interactions among the mother, the fetus,

Advances in Organ Biology
Volume 1, pages 121–152.

ISBN: 1-55938-639-8

and the placenta. Progesterone and estrogens, produced by the feto-placental unit, promote and hinder, respectively, the maintenance of the pregnancy. In several species including the sheep, the maturation of the fetal hypothalamo-pituitary-adrenal axis at term results in elevated fetal cortisol level that causes an increase in the ratio of estrogens to progesterone in the maternal circulation, initiating parturition. Local and diurnal, but not systemic and sustained, increases of the estrogen:progesterone ratio may have a similar function in primates including women. Prostaglandins, synthesized in the gestational tissues may facilitate myometrial activity, cervical maturation, and membrane rupture and thus be of crucial importance as paracrine factors effecting labor onset. Locally produced as well as pituitary oxytocin is a powerful stimulant of myometrial contractions in late pregnancy. Proinflammatory cytokines are likely responsible for early labor in the setting of intrauterine infection. An elevated TH2:TH1 cytokine ratio is part of the mechanism responsible for the immune tolerance of the fetal allograft by the mother. The trophoblast produces a diversity of hormonal peptides and proteins, such as hCG, CRH, GnRH, ACTH, and chorionic somatomammotropin, the function of which is unclear at late gestation. Endothelin, relaxin, prolactin, catecholamines, and growth factors are also present in the late pregnant uterus, but their roles in labor onset remain to be established.

I. INTRODUCTION

Endocrine systems in pregnancy can be considered to have two distinct functions: (a) homeostatic regulation to ensure that the nutritional and oxygen supply to the developing conceptus is sufficient to support optimal growth and development; and (b) maintenance of the pregnant uterus in a state of relative quiescence throughout pregnancy to allow fetal development and maturation and then to cause birth at the time optimal for extrauterine existence of the newborn. This chapter will deal predominantly with the latter function—the maintenance of pregnancy and regulation of the timing of parturition. Disorders of these physiologic mechanisms constitute the major problem in modern obstetrics. Although there is increasing understanding of these mechanisms in some animal species, it is becoming clear that there are many differences between these species and humans. Where possible, we shall provide information from human experimentation. However, there remain many gaps in our understanding of the regulation of human parturition. What is becoming clearer is that many important processes are not occurring in a classic endocrine fashion but rather as *paracrine* events where products from one cell or tissue type are secreted to influence activity in adjacent cells or tissues. Such processes may not be reflected in circulating blood and are therefore very difficult to study.

Potentially important components of intrauterine paracrine systems include the *human fetal membranes*—the *amnion* and the *chorion*. Human implantation occurs as an interstitial process. The fetus and amniotic fluid are contained within the amnionic membrane. This membrane is composed of a single layer of cuboidal epithelial cells with an underlying stroma of very tough collagenous tissue and a few fibroblasts. Human amnion produces large quantities of prostaglandins that generally are thought to play some role in parturition (Challis and Olson, 1988). As the conceptus grows, the amniotic cavity expands and takes with it the overlying layer of trophoblast tissue which becomes the chorionic membrane. This membrane is several cell layers thick and as it stretches around the expanding conceptus, it becomes avascular. The cytotrophoblast cells of this membrane are functionally distinct from the placental cytotrophoblasts. The membrane also contains cells of mesenchymal origin which have not been well characterized. Human chorion may be involved in production of several paracrine factors, particularly steroids but also prostaglandins and peptides. However, it may be more important as a site of metabolism of hormones, particularly steroids and prostaglandins.

The endometrium of human pregnancy is termed *decidua*. This is a complex tissue situated at the interface of the fetal–maternal communication system. It is composed of modified glandular and stromal cells of the endometrium as well as abundant vascular tissue. At term, almost one-half of the decidua is composed of bone marrow-derived cells including macrophages, lymphocytes, and neutrophils. These cells, usually associated with the immune system, produce several *cytokines* that may be important in mechanisms regulating myometrial contractility. The decidua also produces a host of other hormones, including proteins, steroids, and prostaglandins, that may influence the timing of labor onset.

Human parturition can be considered as a sequence of five distinct but likely interrelated events (Figure 1).

(A) The *activation* phase where the uterus is prepared to give maximal response to a stimulant. This phase includes such processes as synthesis of oxytocin receptors and assembly of myometrial gap junctions. It may occur over several days prior to labor onset.

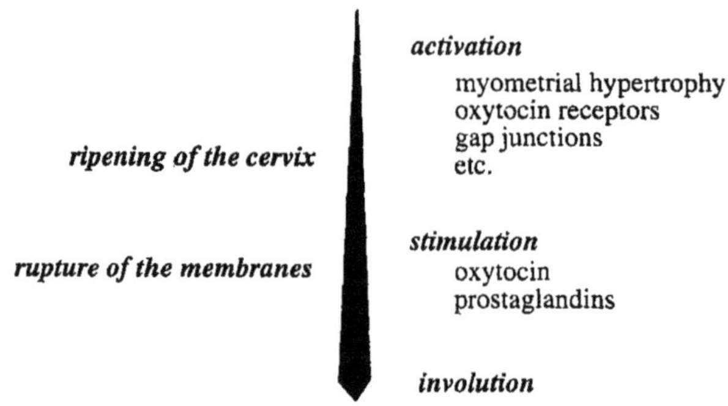

Figure 1. Parturition involves a number of discrete events that occur in an orderly sequence. Pregnancy complications occur when this sequence is delayed or prematurely activated.

(B) *Ripening of the cervix* where the collagen matrix of the cervix is broken down and the water content increased. This renders the cervix soft and pliable and therefore easily effaced and dilated.

(C) *Rupture of the membranes* usually occurs prior to delivery and often heralds the onset of labor. The mechanisms regulating degradation of the amnio-chorionic membrane are not well understood.

(D) The *stimulation* phase characterized by short, frequent high amplitude contractions causing delivery of the fetus. It is not clear whether this phase is caused by a discrete stimulant(s) or whether it represents escape from the mechanisms maintaining uterine quiescence prior to labor onset.

(E) The involution phase where the uterine changes regress and the uterus returns to its non-pregnant state.

A large number of hormones likely play important roles in these processes. The major components of these hormonal regulatory systems will be discussed in this chapter.

II. STEROIDS

In almost all species, parturition is accompanied by a decline in the production and maternal plasma concentrations of progesterone. In most, there is a concomitant rise in estrogen. In sheep, the most extensively studied animal model of parturition, estrogen, and progesterone are produced by the placenta. The key event triggering parturition is maturation of the fetal hypothalamo-pituitary-adrenal axis. The resultant increase in fetal cortisol secretion induces transcription of the 17-hydroxylase/17,20 desmolase gene in the placenta. The increasing activity of this enzyme causes metabolism of progesterone leading to a decrease in maternal plasma concentrations. The progesterone metabolites are estrogen precursors which are subsequently converted in the placenta to estrogen. The increasing estrogen/progesterone ratio leads to several events including synthesis of gap junctions which couple myometrial cells electrically and metabolically, enabling powerful, coordinated contractions. The changing steroid milieu also stimulates production of oxytocin receptors and stimulatory prostaglandins. Thus, the changes in estrogen and progesterone are likely to be responsible for both activation and stimulation of the uterus culminating in the onset of labor in the sheep (Figure 2). The presence of defects causing failure of the fetal hypothalamo-pituitary-adrenal axis, or administration of exogenous progesterone to the ewe will postpone labor onset indefinitely.

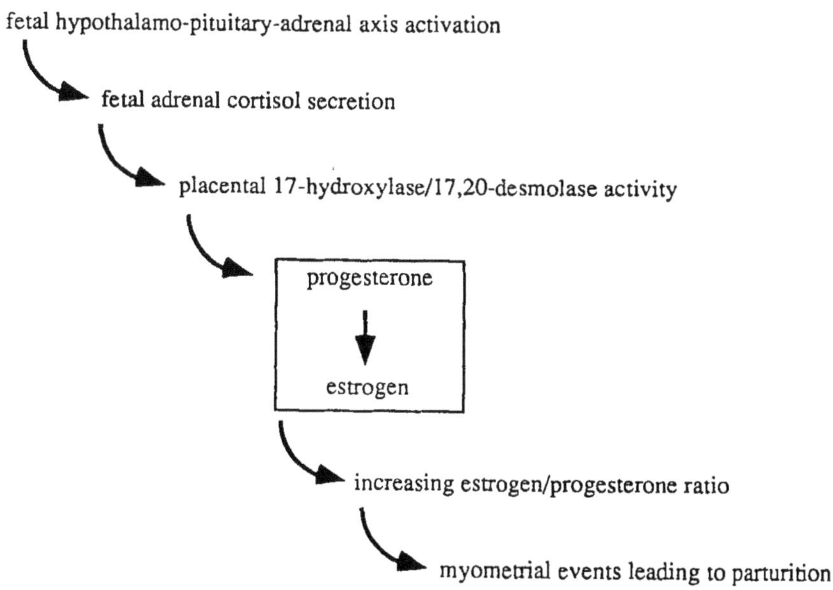

Figure 2. The sequence of endocrine events leading to parturition in the sheep. Maturation of the fetal hypothalamo-pituitary-adrenal axis causes an increase in fetal cortisol which induces synthesis of placental enzymes that result in metabolism of progesterone through to estrogens. The increasing estrogen/progesterone ratio leads to changes in the pregnant uterus that culminate in labor.

Estrogen production in human pregnancy involves an interaction between the fetus and placenta—the *feto-placental unit* (Figure 3). Pregnenolone is synthesized predominantly in the fetal adrenal from LDL-cholesterol in the fetal circulation. The pregnenolone is converted in the fetal adrenal gland to the androgenic estrogen precursor dehydroepiandrosterone sulfate (DHAS). An hydroxyl moiety is attached by fetal hepatic 16-hydoxylase and the resultant 16OH-DHAS is converted to the estrogen estriol in the placenta. Estriol is secreted predominantly into the maternal circulation where it is extensively conjugated to sulfates and glucuronides. This conjugation prolongs its half-life such that estriol accounts for approximately 90% of total circulating estrogens in human pregnancy. Its physiologic role is unknown. Estriol has less biological activity and the concentration of the unconjugated form is less than the major non-pregnancy estrogen, estradiol. The high estrogen levels may be important in maintaining uterine blood flow. Maternal serum or urinary estrogen levels formerly were measured as an index of

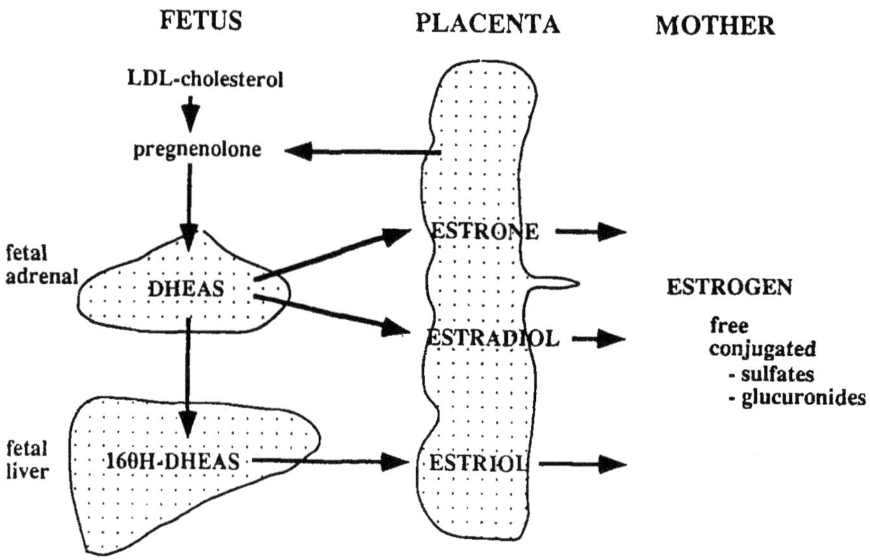

Figure 3. Estrogen synthesis in the human feto-placental unit. Pregnenolone derived largely from fetal LDL-cholesterol is converted to dehydroepiandrosterone sulfate (DHAS) in the fetal adrenal. DHAS is converted in the placenta to estrone or estradiol or is 16-hydroxylated in the fetal liver for subsequent conversion in the placenta to estriol.

feto-placental function. This has largely been abandoned because of the high degree of variability, both within and between patients, and the resultant insensitivity and non-specificity in distinguishing normal from abnormal.

Estrone and estradiol also are synthesized in large quantities in human pregnancy. The androgenic precursors for these estrogens are derived in approximately equal amounts from both the fetal and maternal adrenal glands. Unlike estriol, they do not require the 16-hydroxylation step which is predominantly found in the fetal liver. Thus, their production rates do not reflect function of the feto-placental unit to the extent that production of estriol does.

As in the sheep, progesterone production in late pregnancy occurs in the placenta using maternal cholesterol as substrate. As with estrogen, production rates are much higher than in animal models. In early pregnancy, progesterone is predominantly produced in the corpus luteum of the ovary and removal of the corpus luteum before six weeks menstrual age will result in abortion. However, exogenous progesterone will not delay parturition at term. Progesterone receptor blockers will increase human myometrial contractility. There is an uneasy consensus based on precious little human experimental evidence that, as in animal models, progesterone is likely to be important in maintaining uterine quiescence during human pregnancy.

In contrast to the ewe, the role of estrogen and progesterone in human parturition is controversial. Maternal plasma levels of estrogen and progesterone do not change significantly prior to parturition (Figure 4). However, it is well documented that there are significant diurnal variations in the concentrations of estrogen and progesterone in maternal plasma causing significant changes in the estrogen/progesterone ratio. This may correlate with the well described nocturnal peak in myometrial activity observed in the few days preceding human labor. Furthermore, the hypothesis has been proposed that the basic physiologic mechanisms regulating parturition in humans are similar to those in sheep. According to this hypothesis, the change in the estrogen/progesterone ratio in the human occurs within an intrauterine paracrine system involving the fetal membranes and decidua (Mitchell and Challis, 1988). Such a hypothesis is appealing because the critical changes would occur in the tissues that are in most intimate contact with the myometrium and ideally situated to transmit information between the mother and fetus. Human fetal membranes and decidua produce estrogen and progesterone and several changes in enzyme activity that could cause an increase in the local

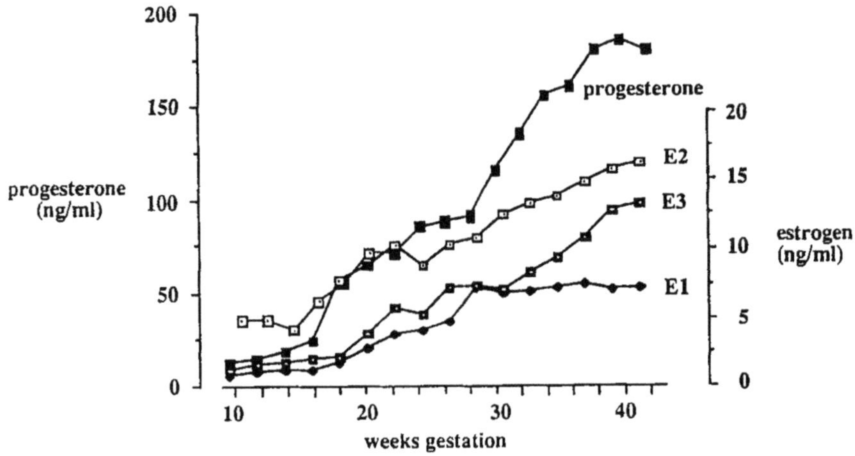

Figure 4. Maternal plasma concentrations of progesterone and unconjugated estrone (E1), estradiol (E2), and estriol (E3) through gestation showing no significant changes at the time of parturition. (Modified from the data of Tulchinsky et al. (1972) with permission.)

estrogen/progesterone ratio occur around the time of parturition. However, a great deal of work remains to prove this hypothesis. Because of the ethical considerations in human experimentation, it may prove a very difficult task to achieve a clear understanding of the role of the sex steroids in human parturition.

As in the sheep, glucocorticoids are important in human fetal maturation, particularly in the production of pulmonary surfactant and biophysical maturation of the fetal lung. However, fetal cortisol does not appear to play an important role in the regulation of parturition. Administration of high doses of glucocorticoids to accelerate fetal lung maturation in women with threatened preterm delivery does not increase the incidence of preterm birth. Fetal anomalies that result in fetal adrenal agenesis do not cause indefinite prolongation of pregnancy as in the sheep. Unlike the sheep, the human placenta freely transports cortisol in either direction. Indeed, the diurnal rhythms in maternal estriol concentrations probably reflect the influence of the maternal adrenal rhythm in cortisol transported across the placenta to cause a reverse rhythm in fetal adrenal estrogen precursor production. Maternal total and free cortisol levels are increased in pregnancy perhaps because its metabolic effects help to ensure adequate nutritional supply to the fetus.

III. GONADOTROPHINS

The human placenta produces large quantities of human chorionic gonadotrophin (hCG). This glycoprotein hormone has biologic similarity to luteinizing hormone (LH) from which it differs biochemically by having a slightly longer β-subunit. It also is slightly more glycosylated, giving it a longer metabolic half-life and greater biologic activity than LH. The major role for hCG appears to be the *rescue of the corpus luteum*. Under maternal pituitary gonadotrophin control, the corpus luteum spontaneously regresses after a life of approximately 14 days. At this time, placental hCG takes over support of the corpus lutem to ensure continued progesterone production and pregnancy survival. Maternal plasma concentrations of hCG peak at approximately 10 weeks of gestation but significant quantities of hCG remain in the maternal circulation throughout gestation. Interestingly, a placental form of LH releasing hormone termed human chorionic gonadotrophin releasing hormone (hCGnRH) has been documented. It is unclear what role this hormone or other regulatory factors have in controlling hCG synthesis or release. There is no good evidence to support a significant role for hCG in steroid production in late pregnancy nor in the mechanism of parturition.

The placenta also produces large quantities of placental lactogen or human chorionic somatomammotrophin (hCS). Although in some species this hormone demonstrates gonadotrophic activity, there is no evidence that this is a significant function in human pregnancy.

IV. OXYTOCIN

Oxytocin has been used pharmacologically since the early twentieth century to stimulate uterine contractions in pregnant women. This nonapeptide is synthesized in the paraventricular and supraoptic nuclei of the hypothalamus. It is transported down axons bound to its carrier protein neurophysin I and stored in the posterior pituitary. Prior to secretion, it is cleaved from neurophysin I, then released into the maternal circulation as the active molecule. It has very specific and potent stimulatory activity on myometrial cells which contain specific receptors. These receptors increase in number in both the myometrium and decidua prior to human labor. Despite its obvious suitability as a physiologic regulator of human parturition, most investigators have discarded the notion that oxytocin has a key role in determining the time of human labor onset. These conclusions have been based on several pieces of scientific evidence. Several investigators have been unable to detect an increase in maternal plasma oxytocin concentrations until labor is well established. Even then, maximal uterine activity occurs when plasma levels are two or three orders of magnitude lower than the K_d of the myometrial oxytocin receptor. Additionally, human pregnancy plasma contains a very active cysteine aminopeptidase, termed *oxytocinase*, that rapidly metabolizes circulating oxytocin. Furthermore, women with posterior pituitary dysfunction usually begin labor at the normal time. Although the fetus synthesizes oxytocin, there appears to be no way that it can escape placental metabolism to reach the myometrium intact. All of these findings suggest it is unlikely for pituitary oxytocin from either mother or fetus to be the key trigger to the stimulation phase of labor.

However, a recent finding could rationalize a role for oxytocin in human labor (see Hirst et al., 1993). Synthesis of mRNA for oxytocin was demonstrated in maternal decidua as well as in amnion and chorion in lesser amounts. The levels of decidual oxytocin mRNA increased significantly around the time of labor. These findings were confirmed in the rat. This supports the hypothesis that oxytocin also could be involved in the paracrine network within fetal membranes and that this network could regulate the timing of the onset of human labor. In addition, sensitive radioimmunoassays have been used to measure small but significant increases in maternal plasma oxytocin levels associated with the nocturnal uterine activity that precedes labor. Oxytocin antagonists have been synthesized and these will block the nocturnal uterine contractions in subhuman primates. These new data suggest that oxytocin may indeed play a significant role in the initiation of human parturition by participation in an intrauterine paracrine network.

V. CORTICOTROPIN RELEASING HORMONE (CRH)

The placenta and the other gestational tissues produce a variety of hypophysiotropic hormones such as thyrotropin, gonadotropin, and growth hormone-releasing hormones, somatostatin, and CRH (Waddel, 1993). Placental CRH has perhaps been the most extensively studied neuropeptide synthesized by the gestational tissues. Due to increasing placental production, the concentration of CRH in the maternal and fetal circulation rises markedly (>25-fold) during the last one-third of gestation, and increases further at labor. Preterm labor is often associated with an earlier rise in plasma CRH levels. Since the activation of the fetal hypothalamo-pituitary-adrenal axis has a well established role in the initiation of parturition in several non-primate species (see other chapters in this volume), it was suggested that an analogous "hormonal axis" might be operating in the primate placenta and contribute to changes in hormone levels eventually leading to the onset of labor (Challis and Hooper, 1989). Indeed, CRH has been shown to stimulate the production of ACTH by the trophoblastic tissue, and the placenta appears to provide trophic support for the maternal as well as the fetal adrenal. *In vitro* studies have indicated that placental CRH production is stimulated by glucocorticoids, implying that a feed-forward cascade exists in the mother and the fetus, which may be responsible for the increases of plasma CRH, ACTH, and cortisol levels as gestation advances. The significance of this putative positive feedback loop in the establishment, maintenance, or termination of primate pregnancy is unclear. CRH was reported to have a direct stimulating action on the myometrium, and was found to enhance the prostaglandin production of gestational tissues *in vitro*. Although these actions might provide a basis to postulate a major role for placental CRH, ACTH, or cortisol in primate parturition, definitive proof for such a function *in vivo* has yet to be provided. Even more work is needed to establish the significance of the other hypophysiotropic factors produced by the placenta.

VI. PROSTAGLANDINS

Prostaglandins are a group of locally acting regulatory factors derived from arachidonic acid, a polyunsaturated fatty acid containing 20 carbon atoms. Arachidonic acid is an essential component of the phospholipids, which are building blocks of membrane structures in the cells; therefore it is present in every cell of the body. The normal metabolic turnover of the phospholipids involves the continuous release and reincorporation of the fatty acid components. This results in a small amount of free arachidonate being present in the cell all the time. However, not all the released arachidonate is incorporated back into lipids; a portion of it is acted upon by various enzymes. One such enzyme is cyclooxygenase, which attaches oxygen molecules to the fatty acid at specific positions. The resulting oxygenated derivatives are further modified by other enzymes in cascade-like sequences of reactions, giving rise to a host of products termed prostaglandins and thromboxanes, many of which have strong biological effects. Figure 5 shows the cascade leading to the formation of prostaglandins and thromboxanes. Some of the biologically active members of the group, such as prostacyclin and thromboxane A_2, are

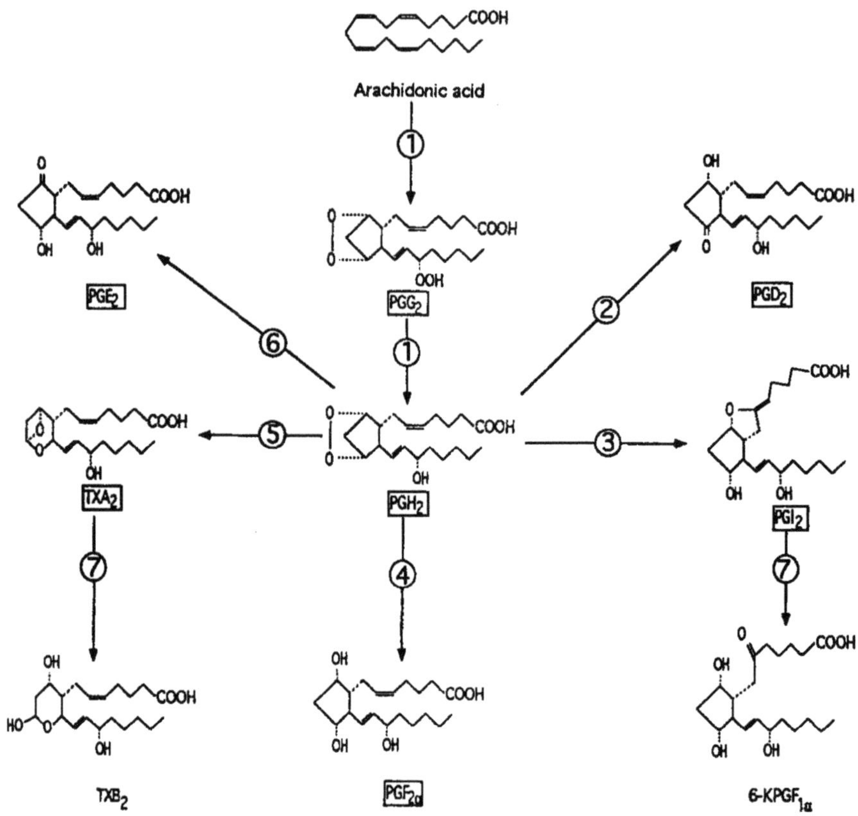

Figure 5. The conversion of arachidonic acid to prostaglandins and thromboxanes. Arachidonic acid is oxygenated to the prostaglandin endoperoxides prostaglandin G_2 (PGG_2) and prostaglandin H_2 (PGH_2) in sequential reactions by the enzyme cyclooxygenase (1). PGH_2 is metabolized further to prostaglandin D_2 (PGD_2) and prostaglandin E_2 (PGE_2) by enzymes with PGD-synthase (2) and PGE-synthase (6) activities, respectively. Prostaglandin $F_{2\alpha}$ ($PGF_{2\alpha}$) is formed from PGH_2 by prostaglandin endoperoxide reductase (4). Prostacyclin (PGI_2) and thromboxane A_2 (TXA_2) are generated from PGH_2 by prostacyclin synthase (3) and thromboxane synthase (5), respectively. Non-enzymatic reactions (7) result in the formation of 6-ketoprostaglandin $F_{1\alpha}$ (6-$KPGF_{1\alpha}$) from PGI_2 and thromboxane B_2 (TXB_2) from TXA_2. Boxes mark the compounds with biologic activities.

chemically unstable; others are rapidly inactivated by enzymes. Because of the ubiquitous presence of their precursors, arachidonic acid and oxygen, and their instability, these compounds are ideally suited to their role, which is the short-term local regulation of a number of physiological functions. Among the varied roles of prostaglandins and thromboxanes are the control of vascular tone, hemostasis, cytoprotection, regulation of fetal sleep/wake states, kidney function, thermoregulation, inflammation, cell proliferation, and many others. An excellent overview of arachidonate release and metabolism to prostaglandins and thromboxanes has been published recently by Smith et al. (1991).

One of the earliest recognized effects of prostaglandins was the stimulation of myometrial contractions. This action, observed experimentally in animal models, was initially related to the fact that the semen contains exceedingly high levels of prostaglandins, and was thought to indicate a role of these compounds in fertility. Later studies revealed, however, that prostaglandins are powerful stimulators of the myometrium, and will terminate pregnancy in a number of species any time during gestation.

In humans, three lines of evidence suggest that prostaglandins are key physiological regulators of myometrial activity in late pregnancy and at parturition (Challis and Olson, 1988; Keirse, 1990). First, the administration of prostaglandin E_2 or $F_{2\alpha}$ to pregnant women induces labor. The prostaglandins not only stimulate the uterus to contract, but also cause the uterine cervix to soften and efface, a process essential for natural birth to take place. Second, prostaglandin E_2 and $F_{2\alpha}$ are synthesized by the intrauterine tissues, and this natural prostaglandin production is higher after the spontaneous onset of labor than before it. In agreement with this, prostaglandins and their precursor, arachidonic acid, accumulate in the amniotic fluid as labor progresses. The maternal plasma concentration and the urinary excretion of the metabolic product of prostaglandin $F_{2\alpha}$ also increase at labor, indicating the enhanced production of the endogenous prostaglandins. Third, it was observed that the length of pregnancy as well as the duration of labor significantly increased in women who were treated with blockers of prostaglandin synthesis to alleviate the symptoms of chronic inflammatory diseases. Moreover, the administration of inhibitors of prostaglandin synthesis slowed down or stopped completely preterm labor. These findings firmly established prostaglandins as the principal factors controlling human birth.

Subsequent investigations were aimed at determining the exact mechanism of prostaglandin action in the pregnant uterus. Important questions were to be answered, such as: Which intrauterine tissues are the sources of the prostaglandins, and which are the targets? What are the primary actions of the prostaglandins in the target tissues? What are the factors which control the synthesis and regulate the inactivation of prostaglandins in the uterine compartment?

Research during recent years made significant progress toward answering these questions (see Olson et al., 1993 for review). The amnion was shown to produce copious amounts of prostaglandin E_2, while the decidua was identified as the main uterine source of prostaglandin $F_{2\alpha}$. It was demonstrated that the amnion synthesizes significantly more prostaglandin E_2 after labor than before the spontaneous onset of parturition. Also, a group of bone marrow-derived cells in the decidua carrying the CD45 surface antigen was shown to synthesize enhanced amounts of prostaglandins at labor. Interestingly, the proportion of these cells increases at term within the heterogeneous population of the decidual cells. At the same time, the chorion exhibits lower activity to synthesize prostanoids, but a very high capacity to inactivate them metabolically.

Administration of prostaglandins to pregnant women induces uterine contractions suggesting that the myometrium is a target of prostaglandin action. Studies with synthetic prostaglandin and thromboxane analogs revealed that the human myometrium contains at least seven types of prostanoid receptors. These receptors bind prostaglandin E_2, $F_{2\alpha}$, D_2, thromboxane A_2, and prostacyclin with different affinities. The receptors are coupled to various signal transduction systems, and are capable of mediating the contraction as well as the relaxation of the tissue. Interestingly, isolated strips of pregnant human myometrium relax when exposed to prostaglandin E_2. This indicates that, at least under *in vitro* conditions, the relaxing effects predominate. *In vitro* studies of prostaglandin transfer across the fetal membranes and the decidua have demonstrated that decidual prostaglandins have unimpeded access to the myometrium because of the anatomical proximity of the two tissues. However, more than 90% of the amniotic prostaglandin E_2 is inactivated in the chorion before reaching the myometrium, suggesting that the primary target of the prostaglandins produced by the amnion membrane is not the myometrium (Olson et al., 1993). Because of these observations, there is a growing consensus among investigators that the traditional view, that labor is initiated by a rise in intrauterine prostaglandin levels which directly stimulate uterine contractions, is too simplistic. It is more likely that prostaglandins act indirectly, or in concert with other agonists to stimulate the uterus to contract. Although the myometrium is undoubtedly one of the chief targets of intrauterine prostaglandins, the regulation of its contractile activity is complex, and not well understood.

The uterine cervix and the amnion membrane are other likely targets of prostaglandins. Both of these structures undergo changes at parturition which result in cervical softening and membrane rupture, respectively.

The cervical changes include the remodeling of the collagen and glycosaminoglycan components of the extracellular matrix. This process is influenced by prostaglandins, especially prostaglandin E_2 (Challis and Olson, 1988; Huszar and Walsh, 1991). The mechanism of prostaglandin action in these tissues is obscure, but it probably involves a local reaction resembling inflammation.

Various *in vitro* studies with perfused placentas, placental explants, cells, or homogenates have shown that the placenta produces measurable amounts of all major products of the prostanoid synthetic pathway (Figure 5; Myatt, 1990). The majority of these compounds are most likely released from the cells of the placental vasculature and participate in the regulation of fetal placental blood flow. The placental trophoblasts are very rich in prostaglandin inactivating enzymes. Similar to the chorion, this high metabolic activity effectively separates the maternal and fetal prostanoid pools.

The regulation of prostaglandin synthesis in human gestational tissues has been extensively studied because of its potential significance in the control of birth. Most of these studies, reviewed recently by Olson et al. (1993), comprised *in vitro* treatments of cells or tissues with agonists, followed by the determination of prostaglandin output or other relevant parameters of prostanoid biosynthesis such as enzyme levels or arachidonate depletion. A variety of natural and synthetic agonists have been found to modulate prostaglandin synthesis in these experiments. Among the steroids, cortisol was demonstrated to inhibit the prostaglandin output of amnion and placental explants, while progesterone or estrogens were not reported to affect prostaglandin production in term human gestational tissues. Oxytocin was shown to stimulate the prostaglandin production of decidual tissue and cultured amnion cells, however, these observations were not confirmed independently (Hirst et al., 1993). ACTH and CRH (corticotropin releasing hormone) increased the prostaglandin output of amnion, chorion, decidua, and placenta at term pregnancy, while GnRH (gonadotropin releasing hormone) stimulated the prostaglandin production of placental explants. Several cytokines, growth factors, and second messenger analogs also affect gestational tissue prostaglandin output; some of these will be discussed in more detail in later sections of this chapter.

Despite the wealth of information concerning the *in vitro* regulation of prostaglandin biosynthesis in the fetal membranes, placenta, and the decidua, little is known about the factors and mechanisms which control intrauterine prostaglandin production *in vivo*. Mechanical stimulation appears to play a role, because artificial rupture of the membranes or distension of the cervix lead to increased prostaglandin levels in the maternal circulation and in the amniotic fluid. However, none of the hormones or paracrine factors which affect the prostanoid output of the gestational tissues *in vitro* have been shown unequivocally to be the physiological regulator(s) of the increasing intrauterine prostaglandin levels observed at labor in women. The criteria to be satisfied by such regulator(s) include: (1) the changing endogenous production preceding or concomitant with the increase of intrauterine prostaglandin levels; (2) the ability to modulate prostaglandin levels upon exogenous administration; and (3) the blockade of the *in vivo* changes of prostaglandin synthesis by antagonists and/or synthesis inhibitors. Additionally, it is necessary to demonstrate that the agonist(s) gain access to the target tissues, the amnion and the decidua, without metabolic inactivation.

These criteria are difficult to establish through experimentation because ethical considerations and technical difficulties often limit the scope of research involving human subjects. Nevertheless, recent data suggest that a group of regulatory substances called inflammatory cytokines may indeed increase prostaglandin levels in the pregnant uterus, at least in certain pathological situations.

VII. CYTOKINES

Cytokines are a group of polypeptide factors which orchestrate the response of an organism to injury, infection, or other effects threatening to upset its homeostasis. They are crucially important in the local as well as the systemic reactions of the acute phase response, and in both the cellular and humoral arms of the immune system. The inflammatory cytokines interleukin 1β (IL1β), tumor necrosis factor α (TNFα), and interleukin 6 (IL6), and the neutrophil chemoattractant cytokine (IL8) accumulate in the amniotic fluid of pregnant women with intrauterine infection and chorioamnionitis. These conditions are often accompanied by preterm labor and elevated amniotic fluid levels of prostaglandins (Romero et al., 1988; Mitchell et al., 1991). Each of the three inflammatory cytokines was shown to stimulate the prostaglandin output of amnion and decidua *in vitro*. The decidua, which contains bone marrow-derived cells, produces TNFα and IL6 in response to bacterial endotoxin and other cytokines such as IL1β. Based on these observations it was proposed that cytokines are important mediators of preterm labor in the case of intrauterine infection and inflammation (Mitchell et al., 1991). It was suggested that the reaction evoked by the bacterial invasion of the uterine cavity leads to increased cytokine levels which stimulate prostaglandin production. The prostaglandins induce labor, and as a result, the fetus is removed from the hostile intrauterine environment. Notably, normal term labor is accompanied by a rise in the number of cytokine producing proinflammatory cells in the decidua. However, no evidence has been provided yet for the involvement of inflammatory cytokines in normal labor.

Another phenomenon where cytokines may be important is the adjustment of the mother's immune system to allow tolerance of the fetal allograft (Colbern and Main, 1991). The suppression of the maternal TH1 lymphocyte dependent (cell-mediated) immunity and the relative prevalence of the TH2 lymphocyte dependent (antibody-mediated) immune response have been established in the pregnant mouse, and the existence of a similar adaptation in human pregnancy is postulated (Wegmann et al., 1993). A maternal immune system with decreased ability to mount a cell-mediated immune response could be advantageous for the maintenance of the pregnancy, because trophoblastic tissue may be susceptible to a cell-mediated immunoreaction. At the same time, enhanced antibody-mediated immunity may provide the newborn with increased protection against infection after birth. The TH1 cell-secreted cytokines interleukin 2, interferon γ, and TNFβ, and the TH2 cell-secreted interleukins IL4, IL5, IL6, and IL10 are most likely involved in this function. The mediating role of prostaglandin E_2 is also conceivable, because this eicosanoid is a strong suppressor of NK (natural killer) cell activity.

VIII. OTHER FACTORS

The vasoactive peptide *endothelin 1* is a potent stimulant of the myometrium. Endothelin 1 is synthesized by amnion cells and is present in term amniotic fluid. Epidermal growth factor (EGF) and IL1 increase the amniotic production of this peptide. However, the chorion contains high levels of the enzyme enkephalinase, which degrades endothelin, and thus prevents it from accessing the myometrium and causing contractions. The physiological function(s) of endothelin in the perinatal period may include the regulation of placental blood flow and the closure of the ductus arteriosus (Mitchell, 1991).

Relaxin is a small polypeptide produced by the myometrium, the gestational tissues and the corpus luteum of pregnancy. Its role in preparing the uterus for labor in animals such as the pig or the rat is well documented (Challis and Olson, 1988). Plasma levels of relaxin increase in these species at term. Relaxin promotes uterine quiescence and cervical maturation before the onset of labor. The plasma of women also contains detectable levels of relaxin throughout gestation. Relaxin has similar effects on the human uterus as in the animal models (Challis and Olson, 1988; Huszar and Walsh, 1991), but its importance in human parturition is not clearly established.

Prolactin is produced exclusively by the decidua in the pregnant human uterus. This hormone was shown to suppress the prostaglandin biosynthesis of the amnion *in vitro*. The influence of prolactin diminishes on the amnion during labor, when the amnion membrane separates from the choriodecidua. This may contribute to the increase of amniotic prostaglandin observed at birth (Tyson et al., 1985).

Catecholamines are present in the amniotic fluid in increasing amounts with advancing gestation. The source of the amniotic fluid catecholamines is most likely the fetal urine. Adrenergic nerve terminals in the uterine cervix may also contribute to catecholamine levels in the uterus. Alpha as well as beta adrenergic receptors are present in the myometrium, the former mediating uterine contractions, the latter mediating relaxation (Huszar and Walsh, 1991). Beta adrenergic agonists are widely used in clinical practice to stop myometrial contractions. Although beta mimetics cause an initial decrease in uterine activity, their efficacy diminishes after several hours of administration probably due to the desensitization of the myometrium. Paradoxically, the human amnion responds to β-adrenergic stimulation with increased arachidonate release and prostaglandin E_2 production (Bleasdale and DiRenzo, 1989). The physiological role of catecholamines in pregnancy and parturition is not understood.

Growth factors are also involved in the regulation of gestational tissue function. The role of epidermal growth factor is characterized best in this respect (reviewed by Olson et al., 1993). EGF and/or transforming growth factor α (TGFα), which is a different polypeptide acting through the same receptor as EGF, were shown to accumulate in the amniotic fluid in late pregnancy and at labor. The sources of the EGF/TGFα bioactivity are most likely the fetal kidneys. EGF stimulates the growth of amnion cells, and promotes endothelin 1 and PGE_2 output of confluent amnion cell cultures. Transforming growth factor β (TGFβ), on the other hand, inhibits the PGE_2 production of amnion cells without affecting mitogenesis. TGFβ was shown recently to antagonize the effect of progesterone on several progesterone responsive genes in the fetal membranes, and was proposed to function as a gene-specific antiprogesterone which might be important in the timing of labor onset.

IX. CONCLUSION

In late gestation, endocrine systems function to maintain uterine quiescence while preparing the myometrium and the cervix for impending labor. At the end of pregnancy, hormonal changes, which most likely involve paracrine interactions within intrauterine tissues, result in the stimulation of myometrial activity culminating in the expulsion of the fetus. Parturition is followed by involution of the uterus to its nonpregnant state.

Progesterone and estrogen levels are high in the maternal plasma in late pregnancy, and, contrary to well studied animal models such as the sheep, remain unchanged before labor. Progesterone is produced by the placenta; estrogen synthesis is the coordinated function of the fetal adrenal and liver and the placenta. Diurnal and/or local variations of the estrogen/progesterone ratio in the uterus possibly influence the contractility of the myometrium and the timing of labor. The human placenta produces large quantities of human chorionic gonadotrophin and placental lactogen, but the function of these hormones in late pregnancy is unknown.

Oxytocin is a powerful stimulant of the myometrium. Oxytocin secreted by the maternal and fetal hypophysis probably does not play a role in the onset of human labor, but locally produced oxytocin may function as a paracrine factor regulating myometrial contractility.

There is strong evidence suggesting that prostaglandins are important factors controlling human parturition. First, prostaglandins induce labor by promoting myometrial contractions as well as cervical ripening. Second, normal labor is associated with increased intrauterine prostaglandin synthesis. Third, inhibition of prostaglandin biosynthesis prolongs gestation and labor. Although a number of hormones were shown to affect the prostaglandin production of the gestational tissues, the *in vivo* regulation of intrauterine prostaglandin synthesis is unclear.

The inflammatory cytokines IL1, TNF, and IL6 accumulate in the amniotic fluid in pregnancies complicated with intrauterine infection and inflammation, and may have a primary role in the mechanism of infection-related preterm labor. Cytokines also are implicated in the modulation of the maternal immune system to tolerate the fetus and the placenta. Other factors such as endothelin, relaxin, catecholamines, and growth factors are also present in the uterus and influence certain aspects of uterine function in late pregnancy, but their physiological roles remain to be established.

REFERENCES

Bleasdale, J.E., & DiRenzo, G.C. (1989). Fetal influence on human parturition. In: *Fetal Medicine Reviews*, Vol. 1 (Rodeck, C., Ed.) pp. 242–288. Blackwell Scient., London.

Challis, J.R.G., & Hooper, S. (1989). Birth: Outcome of a positive cascade. Bail. Clin. Endocrinol. and Metab. 3, 781–793.

Challis, J.R.G., & Olson, D.M. (1988). Parturition. In: *The Physiology of Reproduction* (Knobil, E., Neill, J. et al., Eds.) pp. 2177–2216. Raven Press, Ltd., New York.

Colbern, G.T., & Main, E.K. (1991). Immunology of the maternal-placental interface in normal pregnancy. Semin. Perinatol. 15, 196–205.

Hirst, J.J., Chibbar, R., & Mitchell, B.F. (1993). Role of oxytocin in the regulation of uterine activity during pregnancy and the initiation of labor. Semin. Reprod. Endocrinol. 11, 219–233.

Huszar, G.B., & Walsh, M.P. (1991). Relationship between myometrial and cervical functions in pregnancy and labor. Semin. Perinatol. 15, 97–117.

Keirse, M.J.N.C. (1990). Eicosanoids in human pregnancy and parturition. In: *Eicosanoids in Reproduction* (Mitchell, M.D., Ed.) pp. 199–222. CRC Press Inc., Boca Raton, FL.

Mitchell, B.F., & Challis, J.R.G. (1988). Estrogen and progesterone metabolism in human fetal membranes. In: *The Physiology and Biochemistry of Human Fetal Membranes* (Mitchell, B.F., Ed.) pp. 5–28. Perinatology Press, Ithaca, NY.

Mitchell, M.D. (1991). Endothelin in perinatal biology. Semin. Perinatol. 15, 79–85.

Mitchell, M.D., Branch, D.W., Lundin-Schiller, S., Romero, R., Daynes, R.A., & Dudley, D.J. (1991). Immunologic aspects of preterm labor. Semin. Perinatol. 15, 210–224.

Myatt, L. (1990). Placental biosynthesis metabolism and transport of eicosanoids. In: *Eicosanoids in Reproduction* (Mitchell, M.D., Ed.) pp. 169–197. CRC Press Inc., Boca Raton, FL.

Olson, D.M., Zakar, T., & Mitchell, B.F. (1993). Prostaglandin synthesis regulation by intrauterine tissues. In: *Molecular Aspects of Placental and Fetal Membrane Autacoids* (Rice, G.E., & Brennecke, S.P., Eds.) pp. 56–95. CRC Press Inc., Boca Raton, FL.

Romero, R., Mazor, M., Wu, Y.K., Sirtori, M., Oyarzun, E., Mitchell, M.D., & Hobbins, J.C. (1988). Infection in the pathogenesis of preterm labor. Semin. Perinatol. 12, 262–279.

Smith, W.L., Marnett, L.J., & DeWitt, D.L. (1991). Prostaglandin and thromboxane biosynthesis. Pharmac. Ther. 49, 153–179.

Tulchinsky, D., Hobel, C.Y., Yeager, E., Marshall, Y.R. (1972). Plasma estrone, estradiol, estriol, progesterone, and 17-hydroxyprogesterone in human pregnancy. I. Normal pregnancy. Am. J. Obstet. Gynecol. 112, 1095–1100.

Tyson, J.E., McCoshen, J.A., & Dubin, N.H. (1985). Inhibition of fetal membrane prostaglandin production by prolactin: Relative importance in the initiation of labor. Am. J. Obstet. Gynecol. 151, 1032–1038.

Waddell, B.Y. (1993). The placenta as hypothalmus and pituitary: Possible impact on maternal and fetal adrenal function. Reprod. Fertil. Dev. 5, 479–497.

Wegmann, T.G., Lin, H., Guilbert, L., & Mosmann, T.R. (1993). Bidirectional cytokine interactions in the maternal-fetal relationship: Is successful pregnancy a TH_2 phenomenon? Immunology Today 14, 353–356.

STEROID MODULATION OF MYOMETRIAL STRUCTURE AND FUNCTION DURING PREGNANCY

Charles A. Ducsay and Joon W. Rhee

Advances in Organ Biology
Volume 1, pages 175–193.

ISBN: 1-55938-639-8

ABSTRACT

Throughout the course of gestation, steroid hormones play a key role in regulation of the uterus. Estrogen and progesterone have profound effects on modulation of both structural and functional elements of the myometrium. This chapter will explore the following major areas of steroid-myometrial interaction:

1. Myometrial growth—the dramatic changes induced by estrogen and progesterone during the course of pregnancy;
2. Myometrial excitability/contractility—the role of steroids in regulation of myometrial receptors; and
3. Conductivity—how steroid hormones affect cell-to-cell communication.

I. INTRODUCTION

The successful completion of pregnancy culminating in well timed labor and delivery depends on a series of coordinated processes. This occurs primarily through the interaction of steroid hormones with the myometrium. A considerable quantity of literature has clearly recognized the fall in circulating progesterone concentrations as the harbinger to parturition in a variety of animal species. Excluded from this list, however, are the human and non-human primate. This lack of measurable decline in progesterone at term has been one of the most significant stumbling blocks to unlocking the mysteries of parturition in women.

Due to obvious ethical concerns, we have relied on animal models to gain an understanding of the endocrine, biochemical, and molecular events leading to parturition. All of these changes are responsible for the remarkable transition of the myometrium. It changes from a passive bystander to an active participant in the process of labor and delivery. A key element in this process is a precise interplay of steroids and the myometrium. In this chapter, we will focus on the role of steroid hormones in the regulation of myometrial structure and function during pregnancy.

An abundance of current data suggest that estrogen and progesterone are intimately involved in the maintenance of pregnancy as well as the onset of parturition. The dominance of progesterone throughout the course of pregnancy is associated with a firm, closed cervix, low intracellular calcium levels, and few myometrial gap junctions. All of these factors contribute to the key element of pregnancy maintenance: a quiescent myometrium. At term, there is a marked change in the balance between estrogen and progesterone in the non-primate that stimulates cervical ripening and uterine activity. We will explore how steroid hormones exert such dramatic changes in the following areas of myometrial regulation (Figure 1):

1. Myometrial growth: cell hypertrophy and/or proliferation.
2. Myometrial excitability/contractility: dependent on plasma membrane receptors for contractile agonists.
3. Conductivity: the ability for cell-to-cell communication, that is, propagation of action potentials via electronic coupling through gap junctions.

Before we focus on these areas of hormone-myometrium interaction, it is important to first review how steroid hormones exert their effects at the cellular level. Specifically, we will describe steroid mechanisms, regulation of steroid receptors, and the interaction of estrogen and progesterone at the level of the myometrium.

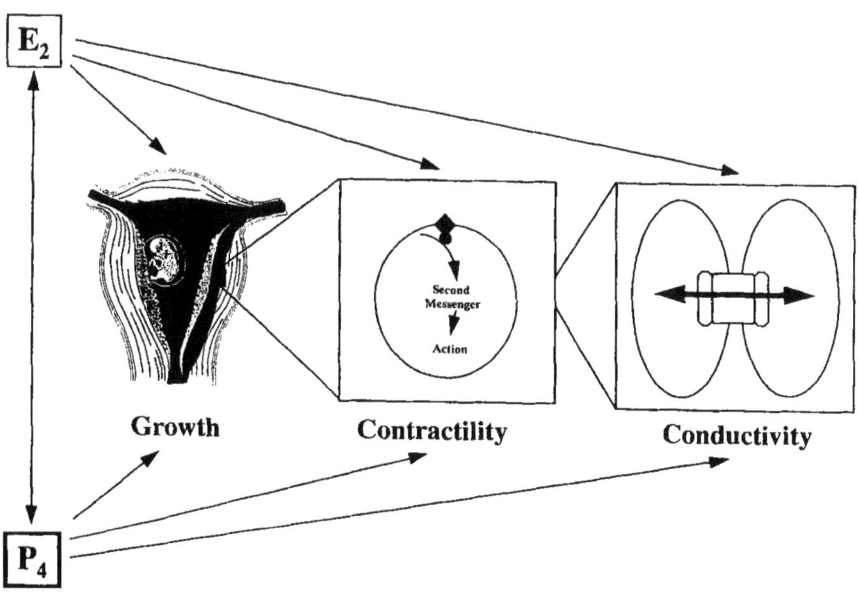

Figure 1.

A. Steroid Mechanisms

The steroid hormones are a large group of molecules all derived from the common sterol precursor, cholesterol. Since steroids are lipid-soluble molecules, they pass through the plasma membrane and bind to specific intracellular receptors. Steroid-receptor complexes (SRC) then enter the nucleus and bind to hormone response elements (HRE) of the regulatory DNA region (Figure 2). Unbound steroid may also go into the nucleus and form complexes inside the nucleus. It is through the interaction of the SRC with the HRE that estrogen and progesterone control a number of physiologic events including hormone production and protein synthesis in the uterus. The DNA binding domain of both estrogen and progesterone receptors has a 65–68 amino acid core region, which contains two essential zinc finger structures, each resulting from the coordination of one Zn^{2+} to four cysteine residues (Gronemeyer, 1991). These structures aid in stabilizing receptor-to-ERE binding. Yet, how the hormone recep-

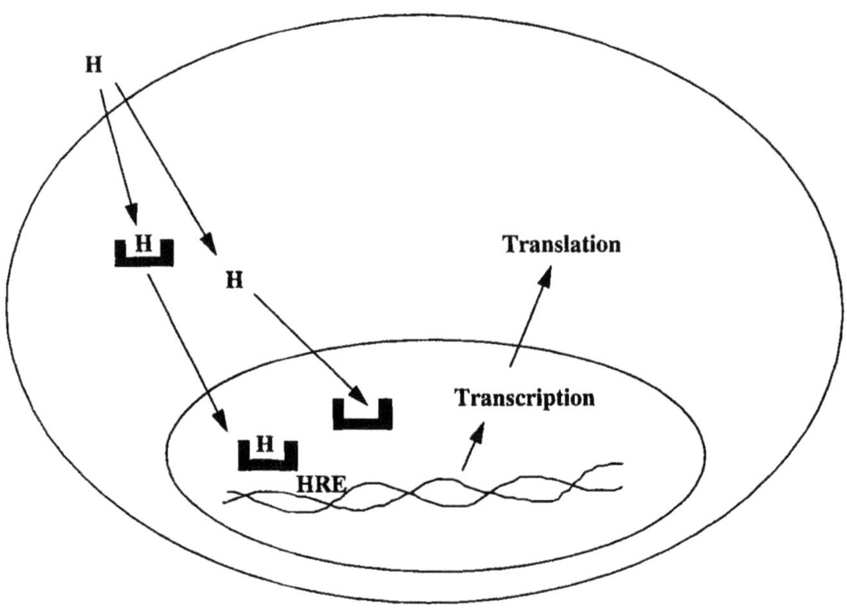

Figure 2. Steroid receptor interaction. Lipid-soluble hormone molecules bind their intracellular receptors. These hormone-receptor complexes dimerize before they enter the nucleus. The complexes then interact with HRE to stimulate transactivation of target genes, followed by transcription and translation. H = Hormone, HRE = Hormone Responsive Element.

tors, once positioned on the right promoter, modulate transcriptional efficiency is not known. Recent findings suggest that modulation of transcription requires functional interactions among the receptor molecules as well as interactions with other essential transcription factors.

B. Regulation of Steroid Receptors

Regulation of steroid receptors is mediated by the steroids themselves. Progesterone receptor induction by estradiol is at the level of transcription because both progesterone receptor gene transcription and mRNA concentrations increase following estradiol treatment. Estradiol helps maintain its own receptor concentration, whereas progesterone inhibits this effect.

In contrast, estradiol and progesterone have also been found to decrease estradiol receptor density: estradiol works via desensitization whereas progesterone acts through antagonism (Okulicz, 1989). This apparent contradiction in estradiol/progesterone interaction is not well understood. It is safe to conclude, however, that the mechanism of self-down-regulation exhibited by both estradiol and progesterone appears to be a desensitization phenomenon when plasma hormone levels are elevated. Why estradiol in appropriate amounts induces an increase of its own receptors and why such induction is not shown by progesterone are subjects of further investigation.

Interaction of Estrogen and Progesterone

How estrogen and progesterone inhibit each other's actions may depend on the organ system. In the myometrium, estradiol stimulates the synthesis of its own receptors and induces the synthesis of cytosolic progesterone receptors. In contrast, progesterone inhibits estradiol nuclear receptor retention, and exerts a negative feedback on its own receptors.

In the rat uterus, cytosolic estrogen receptors exhibit a replenishment phase of 24–72 h following the estrogen treatment. Progesterone inhibits this receptor replenishment, through interaction with the progesterone receptor itself and the degree of inhibition is related to plasma progesterone levels. Low physiologic levels of progesterone are capable of sustained inhibition of occupied nuclear estrogen receptors, thereby decreasing estradiol activity (Okulicz, 1989). In contrast, high plasma progesterone levels depress progesterone receptor concentration by a negative feedback. Additionally, pre-treatment of progesterone diminishes the progesterone antagonism of estrogen-induced uterotrophic responses. High levels of plasma progesterone, therefore, may cause target cells to become unresponsive to progesterone and may play a permissive role in estrogen actions.

These findings are significant for understanding the mechanism of initiation of labor in the primate. Currently, the progesterone withdrawal theory for the initiation of labor is well established for the sheep and rodents, but much debated for the primate because there is no observable decline in plasma progesterone at term. However, increasing levels of progesterone may reach a point where the negative feedback on progesterone receptors occurs, and the withdrawal of progesterone action (receptors) may initiate labor. This hypothesis is strengthened by studies with RU486 blockade of progesterone receptors, functionally decreasing progesterone receptor density in the myometrium. RU486 effectively decreased progesterone inhibition of estrogen receptor retention. However, this decrease in progesterone inhibition of estrogen receptors does not appear to occur in the normal parturient human or non-human primate (Haluska et al., 1990).

It is apparent that the interaction between estrogen and progesterone is important during the course of gestation. We will now discuss the specific roles of these steroid hormones at the level of the myometrium.

II. MYOMETRIAL GROWTH

Following conception, the normal uterus undergoes dramatic changes. It is transformed from an almost solid cavity of 10 cc or less to a relatively thin-walled storage space 500–1000 times its initial capacity by the end of pregnancy. Concomitantly, uterine weight increases about 15-fold, from 70 g to about 1,000 g. The majority of this size increase is the result of stretching and hypertrophy of preexisting smooth muscle cells. At parturition, a single myometrial cell is about 500 μm in length and myometrial hyperplasia is very limited.

During the first few months of pregnancy, when steroid hormones are primarily provided by the corpus luteum, mechanical stretching of the uterus does not play a major role in the process of uterine growth. This is apparent since similar uterine changes can be observed in ectopic pregnancies. However as the products of conception continue to grow, some of the observed uterine changes may be a result of mechanical distention.

Progesterone, in conjunction with estrogen, stimulates uterine growth, as well as causes secretory maturation of the endometrium. However, the cellular mechanisms by which these hormones regulate development of the uterus (both endometrium and myometrium) are poorly understood. Increasing evidence suggests that sex steroids regulate the gene expression of specific proteins that mediate uterine growth during pregnancy. Progesterone induces the expression of uteroglobin and proenkephalin in a number of animal species. The mitogenic effect of estrogen may involve the local production or action of growth factors such as epithelial cell growth factor. These effects occur primarily at the level of the endometrium or decidua. However, other studies have demonstrated the effect of steroid hormones on growth factors in the myometrium. Northern blot analysis showed that the pregnant human myometrium expressed mRNA for both macrophage colony-stimulating factor (MCSF) as well as the c-fms proto-oncogene (the MCSF receptor), while myometrium from non-pregnant women expressed neither mRNA. Interestingly, the mRNAs of both MCSF and the receptor were induced in the myometrium following pseudopregnant therapy with mestranol and norethindrone in a group of non-pregnant patients.

Platelet-derived growth factor (PDGF) has also been implicated in myometrial cell proliferation. The presence of PDGF was confirmed in gestational human myometrium with the amount of A-chain transcript increased during gestation. Further, PDGF polypeptide expression was demonstrated *in situ*. This suggested that PDGF may be associated with myometrial hyperplasia and PDGF may also stimulate smooth muscle cells to increase proliferative capacity. Steroid regulation of PDGF may play a role during pregnancy since PDGF levels in pregnant myometrium are significantly elevated compared to non-pregnant tissue. Additionally, angiotensin-II increases PDGF in smooth muscle cells. Estrogen stimulates plasma renin levels during pregnancy, which in turn, lead to elevated angiotensin-II concentrations. The indirect stimulation of angiotensin-II by estrogen during pregnancy may help explain the elevated PDGF levels observed during gestation. It appears, therefore, that steroid modulation of growth factors may have a profound effect on the dramatic uterine growth observed during the course of pregnancy.

III. MYOMETRIAL EXCITABILITY AND CONTRACTILITY

It is now apparent that both estrogen and progesterone have profound effects on uterine growth. For the uterus to accomplish its goal of pregnancy maintenance followed by well-timed parturition, steroid hormones also play a key role in the regulation of myometrial excitability and contractility. Steroid modulation of these events will be described in terms of modulation of primary contractile agonists and their receptors.

A. Steroid Regulation of Myometrial Oxytocin Receptors

Oxytocin is a potent uterotonic agent with profound effects on myometrial contractility. The early studies of Caldeyro-Barcia and Sereno (1961) showed quite convincingly that uterine sensitivity to oxytocin increased dramatically during the course of gestation. This change is the result of a significant rise in high-affinity low-capacity oxytocin receptors. Compared with the non-pregnant uterus, receptor concentrations increase sixfold by 16 weeks, 80-fold by term, and 200-fold during labor (Fuchs and Fuchs, 1984; Dawood, 1989). Like the adrenergic α_1-receptors, oxytocin (OT) receptors are coupled to phospholipase C via G-protein. This coupling of ligand to receptor stimulates the conversion of phosphatidylinositol-4,5-bis-phosphate (PIP_2) to inositol-1,4,5-trisphosphate (IP_3), which stimulates the release of intracellular calcium and subsequent Ca^{+2} -calmodulin activation of myosin light-chain kinase. Additionally, OT receptor-coupled G-proteins are linked to voltage-gated calcium channels, which regulate the influx of the extracellular calcium.

Estrogen increases OT-receptor density, whereas progesterone decreases the estradiol-induced rise in OT receptor concentration. Incubation of rat uterine tissues in estradiol results in approximately a fivefold increase in the number of OT receptors. This increase is maintained for 48 h, in the presence of estradiol. Additionally, cycloheximide, a protein synthesis inhibitor, blocks the estradiol-induced rise in OT receptor synthesis. However, the reduction of estradiol-induced OT receptor formation by progesterone was not inhibited by cycloheximide. This indicates that while estradiol-induced receptor formation involves the stimulation of protein synthesis, progesterone-mediated OT receptor reduction may be caused by inactivation or degradation of existing OT receptors (Soloff et al., 1983).

Rat myometrium shows a marked increase in the OT receptor density during parturition. This increase is preceded by a proportional rise in estrogen receptor concentration in both cytosolic and nuclear fractions of the myometrium (Alexandrova and Soloff, 1980). In the pregnant rat, progesterone levels fall from day 15 and reaches its nadir on the day of parturition while estradiol progressively rises from day 17 with gestation and reaches its peak on the day of parturition. The decline in plasma progesterone permits estradiol to stimulate the synthesis of estradiol receptors in the myometrium. This increase in estradiol receptors and their occupancy by estradiol promotes the formation of OT receptors, which augments the myometrial response to OT.

The ability to block oxytocin receptors has become a reality with the advent of oxytocin receptor blockers. Both *in vitro* and *in vivo* studies have demonstrated the effectiveness of these agents in inhibiting myometrial contractility. In the clinical setting, it is important to note that oxytocin receptor blockade is relatively ineffective prior to 30 weeks of gestation presumably due to the relatively low number of receptors at this stage of gestation.

Steroid Regulation of Oxytocin Concentrations

In addition to their effects on oxytocin receptors, steroids appear to play a role in the regulation of OT secretory patterns. Fuchs and coworkers (1992) demonstrated a correlation of nocturnal increases in plasma oxytocin with a decrease in the plasma estradiol/progesterone ratio in late human pregnancy. Further, uterine activity in pregnant women exhibits a marked diurnal rhythm (Germain et al., 1993). Marked diurnal variations in uterine activity of pregnant rhesus monkeys and baboons have also been demonstrated (Ducsay et al., 1983; Wilson et al., 1991). It is noteworthy that uterine responsiveness to oxytocin also exhibited a 24 h rhythm, which coincided with the rhythm in uterine activity (Honnebier et al., 1989). Additionally, administration of specific oxytocin antagonists abolished the midnight peak in uterine activity in pregnant rhesus monkeys (Honnebier et al., 1989) and baboons (Wilson et al., 1991). This strongly suggests that oxytocin is responsible for 24 h rhythms in the uterine activity in pregnant primates. Studies with pregnant women (Fuchs et al., 1992) and baboons (Wilson et al., 1991) point to the balance of estradiol and progesterone as an important regulator of uterine responsiveness to oxytocin (Fuchs et al., 1992).

A number of studies have shown that human amnion, chorion, and decidua are capable of producing and metabolizing estrogen and progesterone (Mitchell et al., 1984). The local phenomenon may result in altered estradiol/progesterone ratios in these tissues, not reflected in maternal plasma steroid concentrations. Estrogen and progesterone may therefore exert a paracrine effect on myometrial contractility. Local oxytocin synthesis in human decidua and membranes may play a significant role in the regulation of myometrial contractility at term. Recent use of molecular techniques has provided evidence for oxytocin synthesis by human decidua and membranes (Chibbar et al., 1993). Synthesis of oxytocin mRNA was significantly higher in these tissues after the initiation of spontaneous labor compared with term tissues examined before labor.

Oxytocin Gene Expression

It is well established that estrogen influences hypothalamic oxytocin synthesis, but the regulatory role of steroids in intrauterine oxytocin synthesis is not clearly understood. Preliminary *in vitro* studies, however, have demonstrated a stimulatory effect of estrogen on oxytocin mRNA synthesis in decidua as well as membranes. Alone, progesterone has little effect, but it does inhibit the estrogen response. Oxytocin mRNA has also been isolated from decidua as well as amnion and placenta in the rat. Further, the expression of oxytocin mRNA is greatly enhanced in rat endometrial epithelium during the last four days of pregnancy. This is a period of gestation when estrogen levels are significantly elevated and progesterone levels are decreased (Lefebvre and Farookhi, 1994). Enhanced levels of gene expression were also observed in rats at estrus and in ovariectomized animals undergoing steroid replacement therapy. Thus, the interaction of ovarian steroids in the uterus may represent an important, but probably not exclusive, regulator of OT gene expression. Since uterine epithelial cells possess estrogen-inducible estrogen and progesterone receptors, it has been suggested that the observed steroid effects could be the result of a direct steroid action on uterine epithelial cells.

As described above, the regulation of uterine OT gene expression is another example of the interaction between estrogen and progesterone. Although the estrogen effect may be mediated at least in part by the estrogen response element present close to the OT gene, the mechanisms of progesterone effects on estrogen remain unclear. A glucocorticoid response element has been described in the rat OT gene. In the presence of progesterone receptors, this glucocorticoid response element could mediate the action of progesterone.

The observed steroid effects may also be the result of an integrated response of the uterine endometrium and may involve additional paracrine or autocrine factors. The prostaglandins (PGs) are likely candidates since steroid hormones appear to enhance $PGF_{2\alpha}$ release from endometrium and stimulate luteal secretion of OT. Interestingly, PG release is stimulated by OT itself, and this release is enhanced by progesterone. This raises the possibility that PGs and OT may be linked by a steroid-induced positive feedback loop. However, at present, confirmatory data are lacking.

Taken together, the data discussed above illustrate another mechanism by which steroids can affect myometrial contractility. In this case, the steroid may regulate the paracrine production of oxytocin which, in turn, affects the myometrium. The present data clearly demonstrate that uterine OT gene expression is inducible by steroids. However, the exact mechanism(s) by which this action is exerted remain unclear.

B. Steroid Regulation of Myometrial Adrenoceptors

In the myometrium, the effects of catecholamines are dependent upon the proportion and distribution of the various adrenoceptor subtypes. In turn, the concentration and sensitivity of these receptors appear to be under regulation by steroid hormones, primarily estrogen and progesterone (Roberts et al., 1989). While the majority of data in this area was derived from rodent and rabbit models, the same general principles of regulation appear to apply to the human.

The endogenous adrenergic agonists, norepinephrine and epinephrine, interact with the three principal myometrial receptor subtypes. The cumulative response, therefore, is determined by the relative changes in the products of PIP_2 hydrolysis (α_1-receptors) and changes in cAMP (α_2- and β_2-receptors). These effects increase or decrease the available calcium concentration, respectively, ultimately acting on myosin light chain kinase.

Myometrium contracts in response to norepinephrine, and estrogen treatment increases myometrial contractile sensitivity to norepinephrine. Estrogen treatment results in a dramatic increase in α adrenoceptor concentration while β receptor number is relatively unchanged (Roberts et al., 1989). Although the effects of norepinephrine appear to be mediated through the α_1-receptor subtype, estrogen increases the α_1-adrenergic sensitivity of the rabbit myometrium without changes in α_1-receptor number. It has been suggested that the increased sensitivity is due to a post-receptor effect of estrogen. Specifically, stimulation of α_1-adrenergic receptors activates phosphatidylinositol 4,5 bisphosphate-specific phospholipase C (PIP_2-PLC). This reaction catalyzes the hydrolysis of PIP_2 to form IP_3. IP_3 is responsible for the release of calcium from intracellular stores and the cascade of events leading to contraction of the myometrium. Thus, IP_3 acts as a second messenger, activating calcium responses and the production of IP_3 is enhanced by estrogen treatment.

In contrast to α-adrenoceptors, adrenergic inhibition of contraction is mediated by β-receptors. The response to β-receptor activation is an increase in cAMP and this effect is mediated through a conformational change in the receptor favoring an interaction with a guanyl nucleotide sensitive coupling protein (G_s; Bottari et al., 1983). Like α-receptor stimulation of IP_3, β-mediated stimulation of cAMP and subsequent myometrial relaxation is mediated by alterations in steroid environment. Myometrial cAMP generation is potently stimulated by β-agonists in progesterone treated rabbits but the stimulation disappears following estrogen treatment (Roberts et al., 1989).

C. Steroid Regulation of Prostaglandins

Prostaglandins have been implicated in a number of physiologic roles including blood flow regulation, priming of the uterine cervix during parturition as well as stimulating the decrease of estradiol to progesterone ratio (Alexandrova and Soloff, 1980). PGI_2, for example, increases blood flow by vasodilatation and mediates cervical dilation. $PGF_{2\alpha}$ causes uterine contraction, decreases the estradiol to progesterone ratio, and stimulates OT receptor formation (Alexandrova and Soloff, 1980; Schrey et al., 1988). Suppression of PG synthesis appears to attenuate OT responsiveness and markedly decreases OT receptor concentrations without any change in the binding affinity (Chan and Chen, 1992). Naproxen sodium suppressed both OT receptor and gap junction formation and prolonged gestation and delayed parturition by 24 h or longer.

Prostaglandins increase uterine contractility by stimulating the synthesis of IP_3 just as the α_1-adrenergic agents and OT. Additionally, PGs increase the influx of extracellular calcium. In the rat, $PGF_{2\alpha}$ is the most potent activator of IP_3 on day 0 myometrium, whereas PGE_2 is the strongest stimulus to IP_3 near term. Estradiol stimulates the production of prostaglandins by activating phospholipase A_2 and shifts the receptor population to high-affinity, low-capacity binding-sites. In contrast, progesterone appears to antagonize the effect of estrogen on phospholipase A_2. Further, progesterone treatment decreases the $PGF_{2\alpha}$ receptor concentration without changing affinity.

It has recently become apparent that there is an interaction between oxytocin and prostaglandins. Oxytocin stimulates endometrial PG release, and this release is enhanced by progesterone (Schrey et al., 1988; Steer, 1990). Additionally, oxytocin enhances the basal release of uterine $PGF_{2\alpha}$, but not that of PGE_2. Current data indicate that the phosphoinositol-protein kinase C pathway mediates the oxytocin-induced myometrial contractions as well as the production of PGs. In addition to the stimulatory effects of oxytocin on PG synthesis, there may be a positive

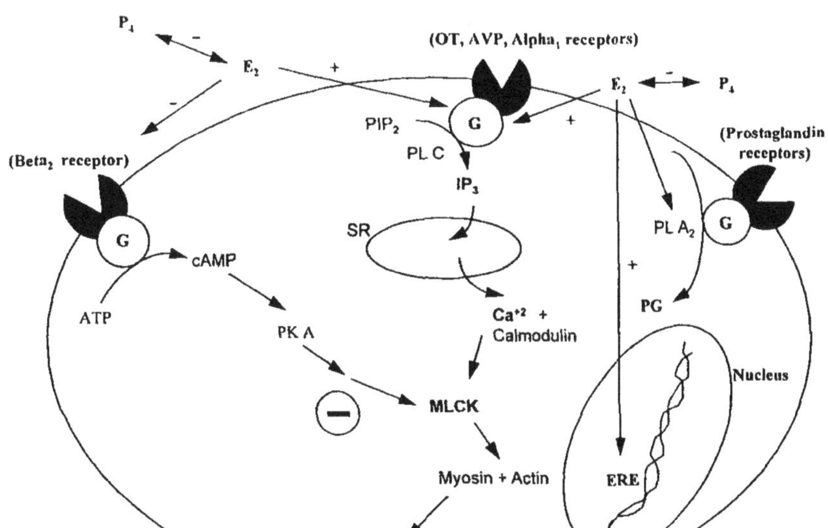

Figure 3. Targets of steroid modulation. Estradiol (E_2) stimulates G-protein coupling of IP_3-generating mechanisms, whereas it blocks cAMP-generating beta adrenoceptor-mediated mechanisms. In general, progesterone (P_4) has the opposite effects. Estradiol also activates PG production through stimulation of PLA_2. ERE = Estrogen Responsive Element, MLCK = Myosin Light-Chain Kinase, PG = Prostaglandin, PLA_2 = Phospholipase A_2, PLC = Phospholipase C, PKA = Protein Kinase A, SR = Sarcoplasmic Reticulum.

feedback cascade within intrauterine tissues that includes oxytocin and PGs as well as the steroid hormones (Hirst et al., 1993).

The effects of estrogen and progesterone on myometrial excitability/contractility are summarized in Figure 3.

IV. MYOMETRIAL CONDUCTIVITY

The early studies of Csapo (1969) predicted that progesterone blocked the conductive properties of myometrial cells, thereby decreasing the contractile capability of the tissue. Recent studies on hormonal regulation of myometrial gap junction formation and function have confirmed this possibility. Gap junctions are structural components of the cell membrane that permit contact dependent, direct cell-to-cell communication. These highly specialized membrane regions have both metabolic and electric signals pass through them and are extremely important in coordinating coupled cellular functions. Due to their role in propagation of action potentials (and perhaps intercellular transfer of second messengers), gap junctions play an integral role in the regulation of myometrial contractility. Our current understanding of the importance of gap junctions in the regulation of uterine activity is largely due to the work of Garfield and coworkers (1977, 1981). Their initial electron microscopic studies first demonstrated the rapid appearance of myometrial gap junctions in rats just prior to and during delivery (Garfield et al., 1977). Subsequent studies demonstrated that an increase in the number and size of junctional complexes was correlated with term and pre-term labor in a number of species, including humans (Garfield and Hyashi, 1981).

Recently, a number of studies have shown that gap junctions exist as a homologous group of proteins that form the intercellular channels of the junctional complex. These proteins, known as connexins, have been identified in a number of tissues including heart, liver, and uterus. Studies utilizing specific antibodies have identified a 43 Kd protein, connexin43 (Cx43), as the dominant junctional protein in myometrium of rats and humans. Northern blot hybridization has also been used to quantify mRNA levels for Cx43 transcription in rabbit myometrium (Beyer et al., 1987; Winterhagen et al., 1991).

A. Regulation of Gap Junctions

All of the methodologies described above have clearly demonstrated the dependence of gap junctions on changes in steroid hormones. In the non-pregnant state and early pregnancy, there is a relative absence of gap junctions. However, at term, there is a dramatic increase in the number of junctional complexes. Estrogen increases gap junction number (Burghardt et al., 1987), while tamoxifen blocks this effect (MacKenzie and Garfield, 1986). Estrogen also appears to stimulate an increase in oxytocin receptors coincident with increased Cx43 (Lye et al., 1993). The increased uterine contractile response to oxytocin at term has been associated with increased gap junction activity.

Studies by Lye and coworkers (1993) demonstrated a number of key points regarding the regulation of gap junctional Cx43 in the rat myometrium. As reported elsewhere (Winterhagen et al., 1991), the levels of myometrial Cx43 increased with gestational age. More importantly, a fourfold increase in Cx43 was observed in ovariectomized rats only 12 h after estrogen administration. This is similar to the time frame we observed in the rabbit which resulted in an increase in contractile sensitivity (Rhee et al., 1992). Further, this time lag is similar to the time between the peak in plasma estradiol concentrations and the nocturnal increase in uterine activity observed in the pregnant rhesus monkey (Matsumoto et al., 1991). Taken together, these data suggest that rhythms in estradiol and/or progesterone influence gap junction formation and/or function over a 24 h period in addition to the dramatic increase observed at term. These graded myometrial responses are well correlated with the ability of the hormone to occupy activated nuclear receptors for a sufficient duration to activate full estrogenic responses.

In contrast to the effects of estrogen, progesterone decreases gap junction number (Puri and Garfield, 1982) and Cx43 gene expression (Winterhagen et al., 1991). The progesterone effect is consistent with the role of this hormone in maintaining uterine quiescence. Prior to parturition in rats and sheep, there is a low estrogen to progesterone ratio (E:P) and a resultant low number of junctional complexes. At term, when the E:P increases dramatically, there is substantial increase in gap junctions in the myometrium. Additional studies in ovariectomized rats have shown that gap junctions induced by estrogen treatment can be eliminated by progesterone treatment.

In the human, this model is not so readily applied. The concept of changing E:P ratios elegantly demonstrated in the sheep is not evident at the end of human pregnancy. Despite no significant changes in the ratio, there is a dramatic and well-defined increase in Cx43 during term labor. In this context, it is important to consider that serum steroid hormone concentrations do not necessarily reflect tissue concentrations, nor do they mirror potential changes in steroid receptors at the tissue levels. Potentially, as suggested by Casey and MacDonald (1993), some form of "local" progesterone withdrawal may occur. This concept would make sense in light of the observed changes in gap junction formation at term. This concept is further strengthened by RU486 data in rats. Despite high progesterone concentrations, blockage of progesterone receptors is capable of stimulating preterm formation of myometrial gap junctions as well as labor and delivery.

Although steroids have profound effects on gap junctions, the question that remains is how is this regulation accomplished? It is assumed that formation of new gap junctions is a function of *de novo* synthesis of channel proteins. Alternatively, estrogen may induce synthesis of mRNA coding for a protein that regulates gap junction expression. Data from Musil et al. (1990) documented that Cx43 is rapidly modified by several post-translational phosphorylations. Recent studies have shown that in viable cells, these phosphorylations are effected by cAMP dependent protein kinase. Further, Dookwah et al. (1992) have demonstrated that cAMP increases gap junction permeability of cultured myometrial cell within minutes. This effect was further enhanced by estradiol treatment. Following uncoupling of the junctional complexes with octanol, permeability was restored with cAMP treatment. These data are in marked contrast to other studies (Sakai et al., 1992) that demonstrated an inhibitory effect of dibutyryl cAMP. Although these differences may be the result of differing methodology, we still have a poor understanding of these regulatory mechanisms. Garfield has proposed that progesterone exerts its effect on junction formation by suppressing a gene regulating connexin synthesis.

Taken together, these data confirm Csapo's hypothesis that progesterone and estrogen play a key role in the regulation of myometrial conductivity and therefore contractility. Although the data are more convincing for experimental animal models, the theory may still apply to the human.

V. CONCLUSIONS

The endocrine events that occur throughout the course of pregnancy are quite remarkable. The specific effects of estrogen and progesterone on the myometrium during this time are equally impressive. As described in this chapter, these steroid hormones have profound effects on myometrial growth. Moreover, there is a well-coordinated effect on the ability of the uterus to respond to a wide range of uterotonins. This is accomplished through regulation of specific receptors. Finally, to allow the uterus to contract purposefully during labor, there is a steroid-mediated increase in myometrial conductivity.

Our efforts to comprehend the key factors involved in endocrine regulation of the myometrium have been hampered by an incomplete understanding of these events in the human and non-human primate models. For obvious ethical reasons, human experimentation is limited. As described in this chapter, a great deal of the information we have regarding steroid regulation of pregnant myometrium has come from animal studies. It may be misleading to directly extrapolate data from lower animal species to fill gaps in our understanding of the endocrine changes that influence myometrial function in the human. However, the molecular and structural alterations that regulate myometrial contractility appear to be similar.

REFERENCES

Alexandrova, M., & Soloff, M.S. (1980). Oxytocin receptors and parturition. I. Control of oxytocin receptor concentration in the rat myometrium at term. Endocrinology 106, 730–735.

Beyer, E.C., Paul, D.L., & Goodenough, D.A. (1987). Connexin 43: A protein from rat heart homologous to a gap junction protein from liver. J. Cell. Biol. 105, 2621–2629.

Bottari, S.P., Vokaer, A., Kaivez, E., Lescrainen, J.P., & Vauquelin, G.P. (1983). Differential regulation of alpha-adrenergic receptor subclasses by gonadal steroids in human myometrium. J. Clin. Endocrinol. Metabol. 57, 937–941.

Burghardt, R.C., Gaddy-Kurten, D., Burghardt, R.L., Kurten, R.C., & Mitchell, P.A. (1987). Gap junction modulation in rat uterus. III. Structure-activity relationships of estrogen receptor-binding ligands on myometrial and serosal cells. Biol. Reprod. 36, 741–751.

Caldeyro-Barcia, R., & Sereno, J.A. (1961). The response of the human uterus to oxytocin throughout pregnancy. In: *Oxytocin* (Caldeyro-Barcia, R., & Heller, H., Eds.) pp. 177–202. Pergamon Press, Oxford.

Casey, M.L., & MacDonald, P.C. (1993). Human parturition: Distinction between the initiation of parturition and the onset of labor. Sem. Reprod. Endocrinol. 11, 272–284.

Chan, W.Y., & Chen, D.L. (1992). Myometrial oxytocin receptors and prostaglandin in the parturition process in the rat. Biol. Reprod. 46, 58–64.

Chibbar, R., Miller, F.D., & Mitchell, B.F. (1993). Synthesis of oxytocin in amnion, chorion, and decidua may influence the timing of human parturition. J. Clin. Invest. 91, 85–192.

Csapo, S. (1969). The four direct regulatory factors of myometrial function. In: *Ciba Foundation Study Group No. 4 on Progesterone: Its Regulatory Effect on the Myometrium* (Wolstenholme, G.E.W., Knight, J., & Churchill, A., Eds.). London.

Dawood, M.Y. (1989). Evolving concepts of oxytocin for induction of labor. Am. J. Perinatol. 6, 167–172.

Dookwah, H.D., Barhoumi, R., Narasimhan, T.R., Safe, S.H., & Burghardt, R.C. (1992). Gap junctions in myometrial cell cultures: Evidence for modulation by cyclic adenosine 3′:5′-monophosphate. Biol. Reprod. 47, 397–407.

Ducsay, C.A., Cook, M.Y., Walsh, S.W., & Novy, M.Y. (1983). Circadian patterns and dexamethasone induced changes in uterine activity in pregnant rhesus monkeys. Am. J. Obstet. Gynecol. 145, 389–396.

Fuchs, A.R., & Fuchs, F. (1984). Endocrinolgy of parturition: A review. Br. J. Obstet. Gynecol. 91, 948–967.

Fuchs, A.R., Behrens, O., & Liu, H.C. (1992). Correlation of nocturnal increase in plasma oxytocin with a decrease in plasma estradiol/progesterone ratio in late pregnancy. Am. J. Obstet. Gynecol. 167, 1559–1563.

Garfield, R.E., Simms, S.M., & Daniel, E.D. (1977). Gap junctions: Their presence and necessity in myometrium during parturition. Science 198, 958–960.

Garfield, R.E., & Hyashi, R.H. (1981). Appearance of gap junctions in the myometrium of women during labor. Am. J. Obstet. Gynecol. 140, 254–260.

Germain, A.M., Valenzuela, G.J., Ivankovic, M., Ducsay, C.A., Gabella, C., & Seron-Ferre, M. (1993). Relationship of circadian rhythms of uterine activity with term and preterm delivery. Am. J. Obstet. Gynecol. 168, 1271–1277.

Gronemeyer, H. (1991). Transcription activation by estrogen and progesterone receptors. Ann. Rev. Genet. 25, 89–123.

Haluska, G.J., West, N.B., Novy, M.J., & Brenner, R.M. (1990). Uterine estrogen receptors are increased by RU486 in late pregnant rhesus macaques but not after spontaneous labor. J. Clin. Endocrinol. Metab. 70, 181–186.

Hirst, J.J., Haluska, G.J., Cook, M.J., & Novy, M.J. (1993). Plasma oxytocin and nocturnal uterine activity: Maternal but not fetal concentrations increase progressively during late pregnancy and delivery in rhesus monkeys. Am. J. Obstet. Gynecol. 169, 415–422.

Honnebier, M.B., Figueroa, J.P., Rivier, J., Vale, W., & Nathanielsz, P.W. (1989). Studies on the role of oxytocin in late pregnancy in the pregnant rhesus monkey: Plasma concentrations of oxytocin in the maternal circulation throughout the 24-h day and the effect of the synthetic oxytocin antagonist [1-beta-Mpa(beta-(CH2)5)1, (Me)Tyr2, Orn8] oxytocin on spontaneous nocturnal myometrial contractions. J. Dev. Physiol. 12, 225–232.

Lefebvre, D.L., Farookhi, R., Giaid, A., Nuculcea, J., & Zingg, H.H. (1994). Uterine oxytocin gene expression. II. Induction by exogenous steroid administration. Endocrinology 134, 2562–2566.

Lye, S.J., Nicholson, B.J., Mascarenhas, M., MacKenzie, L., & Petrocelli, T. (1993). Increased expression of connexin-43 in the rat myometrium during labor is associated with an increase in the plasma estrogen: Progesterone ratio. Endocrinology 132, 2380–2386.

MacKenzie, L.W., & Garfield, R.E. (1986). Effects of taxoxifen citrate and cycloheximide on estradiol induction of rat myometrial gap junctions. Can. J. Physiol. Pharmacol. 64, 703–706.

Matsumoto, T., Hess, D.L., Kaushal, K.M., Valenzuela, G.J., Yellon, S.M., & Ducsay, C.A. (1991). Circadian myometrial and endocrine rhythms in the pregnant rhesus macaque: Effects of constant light and timed melatonin infusions. Am. J. Obstet. Gynecol. 165, 1777–1784.

Mitchell, B.F., Cross, J., Hobkirk, R., & Challis, J.R. (1984). Formation of unconjugated estrogens from estrone sulfate by dispersed cells from human fetal membranes and decidua. J. Clin. Endocrinol. Metab. 58, 845–849.

Musil, L.S., Beyer, E.C., & Goodenough, D.A. (1990). Expression of the gap junction protein connexin43 in embryonic chick lens: Molecular cloning, ultrastructural localization, and post-translational phosphorylation. J. Memb. Biol. 116, 163–175.

Okulicz, W.C. (1989). Temporal effects of progesterone inhibition of occupied nuclear estrogen receptor retention in the rat uterus. J. Endocrinol. 121, 101–107.

Puri, C.P., & Garfield, R.E. (1982). Changes in hormone levels and gap junctions in the rat uterus during pregnancy and parturition in the rat. Biol. Reprod. 27, 967–975.

Rhee, J.W., Long, D.M., Matsumoto, T., & Ducsay, C.A. (1992). Effect of timed estrogen treatment on myometrial contractile sensitivity. Biol. Reprod. Suppl. 1, 46–74.

Roberts, J.M., Riemer, R.K., Bottari, S.P., Wu, Y.Y., & Goldfien, A. (1989). Hormonal regulation of myometrial adrenergic responses: The receptor and beyond. J. Dev. Physiol. 11, 125–134.

Sakai, N., Blennerhassett, M.G., & Garfield, R.E. (1992). Intracellular cAMP concentration modulates gap junction permeability in parturient rat myometrium. Can. J. Physiol. Pharm. 70, 358–364.

Schrey, M.P., Cornford, P.A., Read, A.M., & Steer, P.J. (1988). A role for phosphoinositide hydrolysis in human uterine smooth muscle during parturition. Am. J. Obstet. Gynecol. 159, 964–970.

Soloff, M.S., Fernstrom, M.A., Periyasamy, S., Soloff, S., Baldwin, S., & Wieder, M. (1983). Regulation of oxytocin receptor concentration in rat uterine explants by estrogen and progesterone. Can. J. Biochem. Cell Biol. 61, 625–630.

Steer, P.J. (1990). The endocrinology of parturition in the human. Baillieres Clin. Endocrinol. Metab. 4, 333–349.

Wilson, L., Jr., Parsons, M.T., & Flouret, G. (1991). Inhibition of oxytocin-induced uterine contractions by an oxytocin antagonist in the pregnant baboon. Am. J. Obstet. Gynecol. 165, 456–460.

Winterhagen, E., Stutenkemper, R., Traub, O., Beyer, E., & Willecke, K. (1991). Expression of different connexin genes in rat uterus during decidualization and at term. Eur. J. Cell. Biol. 55, 133–142.

DETERMINANTS OF REPRODUCTIVE MORTALITY AND PRETERM CHILDBIRTH

John A. McCoshen, P. Audrey Fernandes,

Michael L. Boroditsky, and James G. Allardice[†]

[†]deceased

Advances in Organ Biology
Volume 1, pages 195–223.

ISBN: 1-55938-639-8

ABSTRACT

Current global infant and maternal mortality rates are estimated at 66/1,000 and 338/100,000 births per annum respectively. The ranges by country are dramatic and are directly linked to living standards. Historically, mortality rates due to extrinsic factors have been reduced by improving public health policies and social conditions. However, modern medical practice further reduces mortality rates below the thresholds achieved by public health measures by treating intrinsic pathologic factors. Thus, by viewing reproductive mortality from a global perspective, the greatest reductions in infant and maternal deaths can be achieved through public health policies, education, and improved social conditions. On the other hand, developed countries can afford to place greater emphasis on the individual's needs and on personalized obstetrical care. This requires research into the basic mechanisms of pregnancy and parturition, and the development of new therapies to treat a variety of pregnancy related pathologies, particularly that of preterm birth. Moreover, with the current cost effective health care measures being implemented in developed countries, research into the causes of preterm birth and development of truly effective and safe tocolytic treatments will yield substantial savings in this most costly of obstetrical conditions.

I. INTRODUCTION

Women risk significant hazards in bearing children. Unborn babies face even higher risks of mortality and morbidity during the perinatal period.

For a woman there is no greater tragedy than the loss of a child. Yet, while we enjoy the golden age of reproductive medicine, the vast majority of women around the world are deprived of modern health policies and services. Reproductive mortality rates vary widely between world regions and nations, and indeed, even within many nations. In underdeveloped and developing countries, maternal and infant mortality rates are at levels that would not be tolerated in our society. In developed countries, vast improvements in prenatal care and obstetrical services have evolved from a better understanding of the many extrinsic and intrinsic factors that deleteriously impact upon pregnancy outcome. As a result, puerperal mortality rates have declined dramatically, but for reasons that relate more to social and public health measures than to the implementation of high-tech medicine. As yet, the scientific nature of many of the pathologies peculiar to pregnancy and parturition remains a mystery and so continues to evoke the tragedy of reproductive mortality.

Currently, the major obstetrical challenge in developed nations is that of prematurity, or preterm birth, which is the main contributor to infant mortality and morbidity. To some extent it remains an issue of public health, but because rates have not declined over the past few decades, preterm birth is now viewed in the context of a complex of basic pathophysiologies requiring medical solutions. Moreover, while low birthweight infants delivered at only two thirds normal gestation can now be kept alive, the intensive care cost is enormous and, following discharge, survivors suffer higher rates of impairments and major disabilities than term delivered babies. It is the associated mortality and morbidity of prematurity that continue to plague mothers and children and begs for both imaginative and intensified efforts to determine the causes and to effect solutions.

While some advocate that most preterm childbirths reflect adverse social conditions, and may in fact be rising because of increased adolescent pregnancy rates (greater than 25% of U.S. teenage girls get pregnant), all would agree that breakthroughs in elucidating the mechanism of labor is essential in preventing prematurity in afflicted mothers. Except in cases of severe maternal disease, fetal development proceeds best in its natural maternal incubator. Thus, by cracking the mystery of how labor is initiated and sustained, development of more effective and reliable tocolytic agents and regimens than are now available for the prevention of prematurity will yield obvious human benefits including substantial social and economic gains.

In this chapter, we discuss demographics and epidemiologic data on reproductive mortality and preterm childbirth as a purview of the evolution of their solutions at both the public health and scientific levels. Historically the greatest advances in population based health come from improvements in hygiene, nutrition, and public and personal health care policies. However, from the perspective of the mother and fetus at risk for an obstetrical tragedy, the need for fundamental research into the nature of pregnancy and parturition continues if effective clinical interventions are to be developed.

II. REPRODUCTIVE MORTALITY

Reproductive mortality is used here to include maternal deaths associated with pregnancy, fetal death from 20 weeks gestation, puerperal mortality, and mortality of children in their first year of life. Many of the factors contributing to each are common to all and include diseases of heredity, deficiency and infection, malnutrition, maternal age and multiparity, and a number of social and public health deficiencies. Fetal deaths prior to 20 weeks gestation are classified as abortion from either spontaneous (miscarriage) or therapeutic cause. In the United States, the therapeutic abortion rate is between two to threefold greater than the rate of miscarriage. Globally, about 150,000 unwanted pregnancies end in abortion each day (World Health Organization, 1992). While an abortion is a reproductive mortality, miscarriage is unwillful whereas therapeutic abortion is the deliberate termination of pregnancy for reasons that differ radically from those of reproductive mortality from 20 weeks onward. Moreover, reproductive mortality extends back even to preimplantation wastage from biological causes. Thus, our discussion will be confined principally from the twentieth week of gestation to the puerperium, and, in part, to one year postpartum.

A. Historical

Reductions in high rates of infant and maternal mortality and morbidity in North America resulted directly from the accumulation of statistical data. Only in the late nineteenth century did puerperal mortality become a matter of public concern in the newly formed democracy (Spreet, 1980). The first U.S. law requiring compulsory registration of deaths was passed in Massachusetts in 1842, beginning the accumulation of vital statistics that was to be instrumental in addressing medical issues in the

United States. In 1900 the U.S. Bureau of the Census began publishing vital statistics, but not until 1915 did the reliability of birth and death records increase with the establishment of the Birth Registration Area. In that year, the puerperal maternal death rate was 61 per 10,000 live births (compare to a current average of 4/100,000 in Canada; Steinberg, 1989). Among the then top 20 industrialized nations, the United States ranked eleventh in infant mortality and seventeenth in maternal mortality. Maternal mortality in Cincinnati alone was 94 per 10,000 births. Deaths from other causes during this same period declined significantly. For example, mortality per 100,000 from typhoid fever declined from 46 to 17, diphtheria from 97 to 18, tuberculosis from 250 to 147, and pneumonia from 186 to 132. Overall mortality between 1900 and 1915 dropped from 17.2 to 13.2 /1,000. Currently, it is 8.7/1,000 in the United States and 7.4/1,000 in Canada. Worldwide, it is about 10/1,000 with a range of 2–41/1,000 (Steinberg, 1989).

B. Practices of Indigenous Peoples

Clues to controlling puerperal mortality existed among the indigenous peoples of North America who practiced relatively sophisticated obstetrics well before the arrival of the Europeans. Harold Spreet, citing early records of surgeons who collectively spent many years among the American Indians, found little evidence of high maternal and infant mortality (Spreet, 1980). Numerous natural agents including oxytocics and analgesics were employed by native Indians and in many tribes elaborate preparations of the mother preceded the birthing process. In some tribes, parturition huts and midwives, usually older experienced women, facilitated birth. Breast feeding was virtually universal. Family size was relatively small because of prolonged breast feeding, voluntary abstinence, and the use of antifertility herbals that are only now receiving some attention as effective contraceptives. Indeed, the medical practices in general appear to have been well advanced over that of the Europeans. However, following European infiltrations and intermarriages between white men and native women, major obstetrical problems arose. Many native tribes opposed intermarriage, in part, for obstetrical reasons because of an associated increase in infant and maternal mortality that resulted from the delivery of babies too large to pass through the pelvis (Spreet, 1980).

III. GLOBAL INFANT AND MATERNAL MORTALITY

A. Infant Mortality

Infant mortality in some nations is astonishingly high. Rates are highest in Afghanistan and the Western Sahara (>150/1,000 births) and lowest in Iceland and Japan at 4/1,000 (U.S. Bureau of the Census, 1994). The rates correlate inversely to income and positively to the lack of medical services as measured by the number of physicians per 10,000 population (Demographic Yearbook, U.N., 1992). Interestingly, the strongest correlation is to education as measured by literacy rate. Figure 1, based on 180 nations reporting health statistics to the United Nations, displays the relationship of infant mortality according to a quartile distribution of average per capita income. The data, when further segregated by percent literacy (also in quartiles) shows a clear association between income and literacy. Moreover, literacy is the stronger correlate. When the number of physicians per 10,000 population was factored into the analysis, it carried the least strength in the correlation (data not shown). It is apparent that national living standards, especially education, have a direct and positive impact on the mobilization of resources to reduce mortality. Living standards include daily caloric intake, hygiene, and access to health services.

The global birth rate is 26/1,000 population with a range of 7–53 according to nation. Infant mortality is 66/1,000 births with a range of 4–164 by country (Demographic Yearbook, U.N., 1992). The World Health Organization estimates a current annual total of 9.4 million babies who die within their first year of life (*World Health Statistics Annual*, 1994). This represents almost 20% of the 50 million annual deaths around the world. Fully one third of infant mortality occurs perinatally. Under the best of living circumstances, a calculated minimum number of premature births is about 14.5 million babies per annum, but it is probably much higher. As with maternal mortality, prematurity and infant mortality can be drastically cut without direct medical intervention since most of the tragedies of pregnancy occur in the poorer countries. In the United States, as in other developed nations, perinatal, fetal, and neonatal mortality rates steadily declined until about 1980, after which the rates have remained unchanged (Cunningham et al., 1993a) as displayed in Figure 2. The four principal causes of infant mortality in the United States are congenital anomalies (21%), sudden infant death syndrome (15%), preterm birth and low birthweight (11%), and respiratory distress syn-

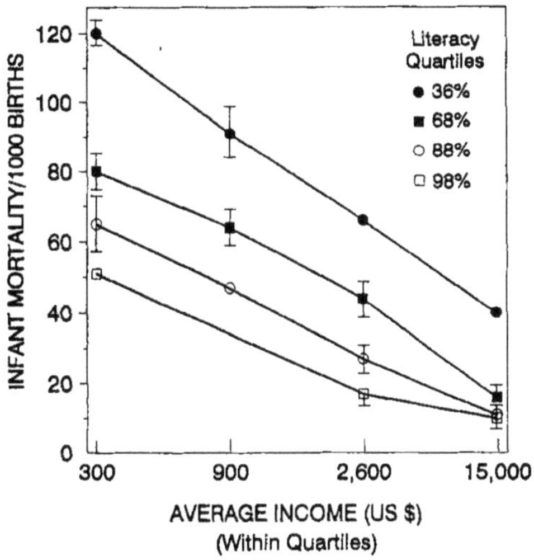

Figure 1. Global infant mortality rates according to average income segregated by quartiles. The inverse relationship is further segregated by percent literacy (also in quartiles). Regardless of mean national income, the highest mortality rates exist for populations with the lowest literacy ($p < 0.0001$) used here as a measure of the educational structure within nations.

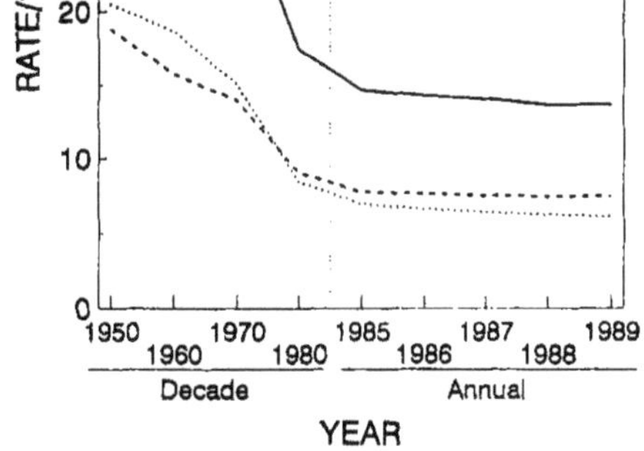

Figure 2. The rate/1,000 births of perinatal, fetal, and neonatal mortality in the United States continued to significantly decline until the 1970s. Since then, there has been no substantial change in mortality except a reduction in neonatal rates to less than that of fetal mortality from 1980. (Based on data from Cunningham et al., 1993a.)

drome (7%). Globally, prematurity is among the top 10 leading causes of childhood death and is associated with an estimated 430,000 deaths per annum (Wright, 1994).

B. Maternal Mortality

Maternal mortality is estimated at greater than one-half million women per year dying during childbirth (about 1,440 per day), most in underdeveloped nations (Steinberg, 1989; Wright, 1994). Of these, 1% occur in developed countries which represent about 20% of the world population. The remaining 99% is distributed within approximately 80% of the world population, representing an average 20-fold higher rate. Based on an estimated 148 million annual births, the global maternal mortality rate is 338/100,000 births versus about 10 or less in the industrialized Western societies. The goal for North America in the 1990s is a reduction to 5/100,000 births (Steinberg, 1989). High maternal mortality is directly linked to public health deficiencies, illiteracy, early marriage and multiparity, and cultural practices. An estimated 35% reduction in maternal mortality could be accomplished if women who desire no more children were able to practice family planning and avoid pregnancy in early and late reproductive ages and high parity that is associated with maternal mortality. Where high standards of living, application of family planning, and availability of good health care services exist, basic pathophysiologies of pregnancy including hemorrhage, infection, unsafe abortion, hypertension, and obstructed labor remain the leading causes of maternal mortality (*World Health Statistics Annual*, 1994). Less than one-half of all pregnancies around the world are delivered in an institution or with professional assistance (World Health Organization, 1992).

IV. WHEN ARE BABIES BORN?

The current estimate of early clinical pregnancy wastage (prior to 20 weeks gestation) following implantation is 13% among young healthy pregnant woman (Cunningham et al., 1993b). The rate increases significantly with maternal age older than 30–34 years (Harlap et al., 1980). Premature birth is defined as birth between 20 weeks gestation to the thirty-sixth completed week. The percentage of total and cumulative births occurring prematurely and at term (37th+ weeks) is compared in Figure 3. Moreover, a comparison of birth weight and percentage sur-

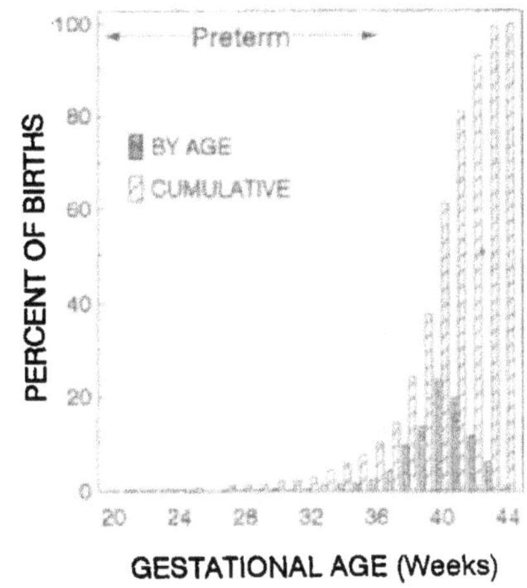

Figure 3. Diagrammatic display of the percentage of total births for each gestational age from 20 weeks onward, and the cumulative percent of births. Most births occur during the fortieth week and about 10% of all births occur before the thirty-seventh week.

vival of babies in the most critical period of prematurity is detailed in Figures 4a and 4b, the latter showing survival as a function of fetal weight. These figures show that the probability of survival approaches 98% by 33 weeks gestation and a birth weight of almost 1,900 grams (Copper et al., 1993). However, a high probability of survival (95%) is found at a birth weight of 1,250 grams (Figure 5). Among 1,262 births by women followed with diagnostic ultrasound at the Women's Hospital

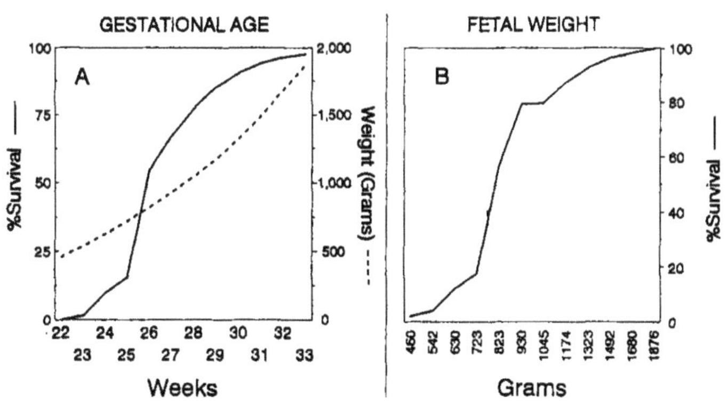

Figure 4. Percent survival (solid lines) of infants born between 22–33 weeks (A) and by birthweight (B) during the same time period. Normal birthweights are plotted (broken line) by age in (A). Survival increases abruptly from 25 weeks or 700+ grams. (U.S. data, based on Cooper et al., 1993.)

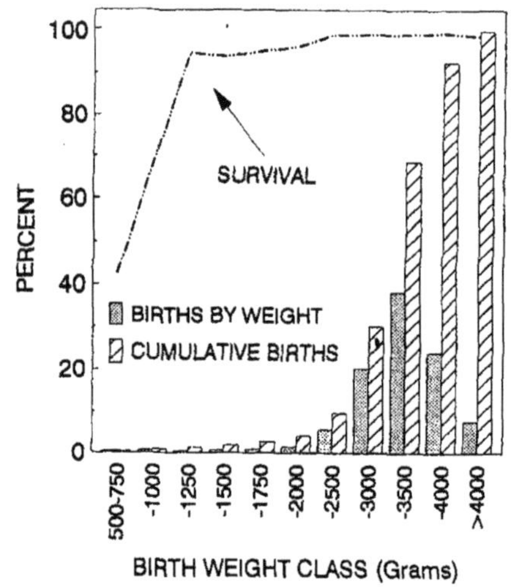

Figure 5. Total and cumulative percentage of births based on classes of fetal weight, and percent survival (dashed line) demonstrating the high survivability of infants of 2,500 grams and greater. Low birthweights are classed as 2,500 grams and less.

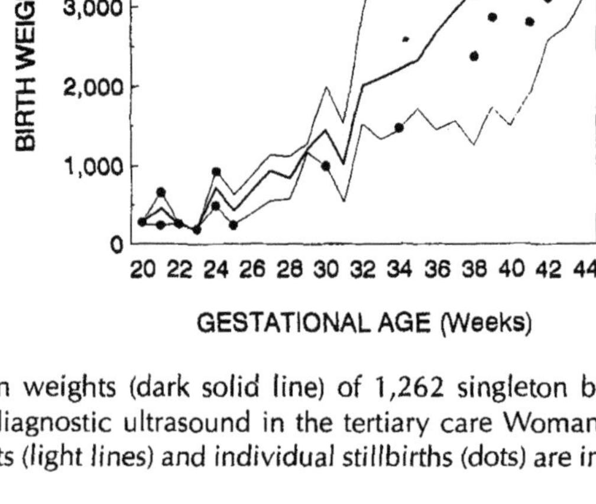

Figure 6. Mean weights (dark solid line) of 1,262 singleton births by women followed with diagnostic ultrasound in the tertiary care Woman's Hospital. The ranges of weights (light lines) and individual stillbirths (dots) are included. See text for details.

in Manitoba, Canada, a tertiary care center, 129 (10.2%) were premature births of which 12 (9.3%) were stillborn. The birth weight versus the gestational age of all live births and stillborn births for both premature and term conditions is displayed in Figure 6. The percentage of stillborn infants was 0.95% between 20 and 44 weeks gestation, and 0.55% were delivered prior to 37 weeks.

V. PRETERM BIRTHS AND DEATHS

About one-third of low birthweight infants (<2,500 grams) may be chronologically mature (Gruenwald, 1963). These infants who are small for their gestational age are considered growth retarded. Fetal growth retardation does not imply risk of spontaneous preterm delivery and the etiologies differ markedly from those of premature labor. Indeed, fetal weight is a function not only of chronological age and pathophysiology, but also it is determined by maternal and familial factors resulting in constitutionally small babies who are otherwise normal and deliver at the appropriate time (Gardosi et al., 1992). Thus, while birthweight is linked to probabilities of infant survival, birthweight data on prematurely born infants is exclusive of the low weights of growth retarded fetuses born at term.

The majority of perinatal deaths occur among infants born preterm. Birth data clarify the highest risk gestational ages and fetal weights for infant survival. The probability of survival is negligible between 20 and 22 weeks, subsequently rising to approximately 15% by 25 weeks. However, the cumulative percentage of births at that time is very low (<2%). Above 25 weeks, the probability of survival increases dramatically with each week, rising to 90% by 30 weeks. Fetuses of 1,000 grams or more have a survival rate of 97%. Certainly for fetuses attaining a weight of 1,500 grams or more, the survival rate closely approaches 100% (Creasy, 1994).

Copper et al. (1993) report that of 33,401 pregnancies, 83% of neonatal deaths resulted at less than 37 weeks gestation and 66% occurred prior to 29 weeks. Among the latter deaths, twin infants were at a three to fourfold higher risk than singletons, and males had almost double the mortality rate of female infants. Noteworthy are reports of significantly higher male karyotypes among spontaneously aborted fetuses in early pregnancy prior to 20 weeks (Kellakumpu-Lehtinen and Pelliniemi, 1984) that may reflect a highly disproportionate rate of conception in favor of the male gender (Brockkov and Kostrava, 1973) but is subsequently compensated for through miscarriage. Overall, it is well known that significantly more males are born than females, the ratio being approximately 106:100 (Hytten, 1982). Thus, gender exclusion appears to continue even up to the twenty-ninth week of pregnancy and may, in part, account for, and reflect, a male component in premature labor onset.

A. Etiologies

Approximately two-thirds of preterm births are of idiopathic origins that provoke labor onset despite intact membranes (30%) or result from preterm rupture of membranes in 35% of cases (Arias and Tomich, 1982). The remaining 35% is associated with other fetal and maternal complications principally including plural pregnancy, polyhydramnios, gross fetal and placental anomalies, fetal death, cervical incompetency, maternal systemic disease, and previous preterm delivery (Savity et al., 1991). Table 1 summarizes the main intrinsic and extrinsic factors associated with prematurity. Currently, preterm spontaneous rupture of membranes, chorioamnionitis, and amniotic fluid infection are receiving considerable attention in premature labor onset (Romero and Major, 1988). Under those conditions, respiratory distress syndrome and neonatal mortality are increased by three and fourfold, respectively (Morales, 1987). Trials of antimicrobial therapy have not as yet conclusively demonstrated

Table 1. Risk Factors Associated with Preterm Birth

Medical	Demographic
Poor Obstetric History	Age
Malformations of Uterus or Cervix	Race
Plural Pregnancy	Socioeconomic Status
Previous Preterm Birth	Marital Status
Fetal Abnormalities	
Placental Anomalies	**Behavioral**
Preterm Membrane Rupture	Smoking
Polyhydramnios	Inadequate Nutrition
Urinary Tract Infection	Excessive Physical Stress
Chorioamnionitis	
Intra-amniotic Infection	**Health Care Services**
Maternal Systemic Disease	Inadequate Prenatal Care
Fetal Death	

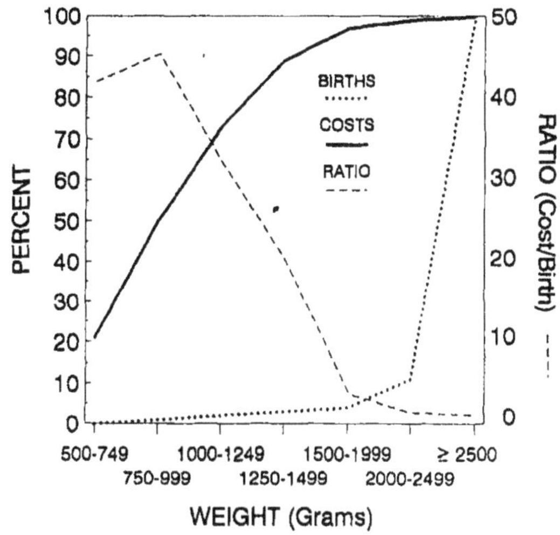

Figure 7. Comparison of percent births by classes of birthweight to intensive care costs up to discharge (percent of expenditures) and the cost/birth ratio demonstrating the consumption of resources of preterm infants.

prolongation of pregnancy or improved infant outcome (Christmas et al., 1992).

As a measure of the degree of care required by premature infants (measured in hospital days), Figure 7 compares the percentage of births by birthweight with corresponding costs, also expressed as a percentage of total expenditures for all births of infants of 500 grams and greater. It is clear that the cost of care for infants of 1,500 grams or less, consumes almost 95% of the monetary expenditures allocated for neonatal care.

VI. HOW MUCH DOES PRETERM BIRTH COST?

The true value of health care cost estimates is realized in the context of the calculated cost of a chronic clinical problem (service) versus the acute cost of solving the problem (research). Cost analyses reveal the extent of a condition because they require population statistics in their calculations that are then compared to the incidence of adverse outcomes within the population being serviced after therapeutic regimens are instituted.

In 1991 the United States had approximately 4,111,000 live births of which 23.5% were cesarean deliveries. The average cost of deliveries was $7,826 for cesarean and $4,720 for vaginal births. The prices include physician and hospital fees. The average hospital stay for all deliveries was 2.8 days (Graves, 1993). The actual cost of the birth of a premature infant is realized not in the delivery costs, but rather in the cost of postpartum medical care that is a function of fetal weight (see Figure 7). In 1984, Walker et al. (1984) performed a cost-benefit analysis of care for low birthweight infants of less than 1,000 grams in the state of Rhode Island. The population of Rhode Island is about 1,000,000 persons. They calculated costs by 100 gram increments from 500 to 999 grams for 247 premature infants and reported the cost per survivor (none survived at 500–599 grams) to be $363,000 at 600–699 grams, $116,000 at 700–799 grams, $101,000 at 800–899 grams, and $41,000 at 900–999 grams. For those 79 surviving, the average cost was $77,600 per baby, and the total cost was more than $6,000,000. Extrapolating those figures to the total number of live births in the United States, that is, about 0.50% are less than 1,000 grams and about 30% survive, the minimum cost of infants less than 1,000 grams is about one-half billion dollars. However, for all premature births, the total cost of care to discharge as well as long-term expenses has been calculated at almost three billion dollars annually (Schwartz, 1989).

In Canada, health care costs differ from the United States because of the economics of our system of universal coverage. In the province of Manitoba, also with a population of 1,000,000 residents, the number of annual births approximates 16,000. Between 1979 and 1989 the average length of stay by birthweight class ranged from 2.5 (>2,500 grams) to 82 (500–999 grams) days. The average number of births of less than 2,500 grams was 728 per annum of which 602 survived to discharge. An annual average of 61 births were 500–999 grams of which 24 survived to discharge. Assuming a hospital per diem of $400, the total hospital cost of all surviving births was more than $24,000,000. By far, the greatest cost exists for birthweights of 500–999 grams ($101,680 per case) as compared to birthweights of 2,500+ grams ($1,000 per case). If by clinically delaying birth by shifting even 20% of low birthweights to the next highest weight class, the averted expenditures equate to more than $1,340,000 in hospital costs alone (Institute of Medicine, 1985; Manitoba Centre for Health Policy and Evaluation Report, 1991; Manitoba Centre for Health Policy and Evaluation Report, 1993). Expenditures by birthweight are compared to infant survival before and after the calculated 20% shift to the next highest weight class in Figure 8. Extrapolating those data to the entire nation with almost 29,000,000 persons (1994 estimate), the hospital savings are approximately $39,000,000 annually.

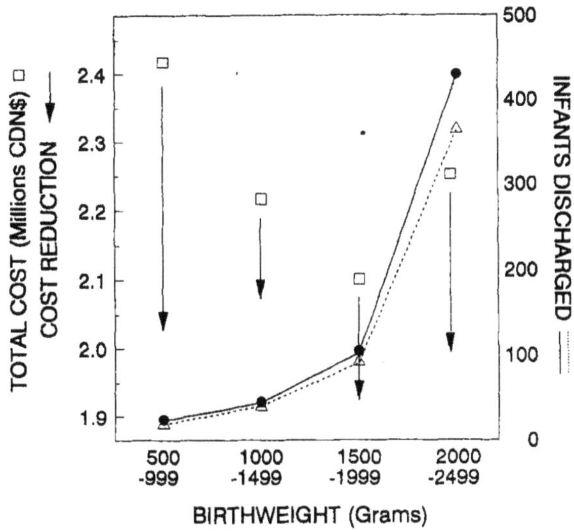

Figure 8. Manitoba demographics (1989) of number of discharged low birthweight infants (solid dots) and hospital costs (open squares), and the calculated effect of a hypothetical shift of 20% of low birthweights to the next higher weight class (open triangles) on the hospital cost reductions (arrows).

Thus, even relatively short-term delay of premature labor carries substantial cost saving, not to mention the associated reduction in infant mortality that would be realized.

VII. WHAT CAUSES PREMATURITY?

Prematurity has been a long-standing cause of perinatal mortality and morbidity. Although increased technology in neonatal care has improved the outcome of preterm birth, the direct causes of prematurity remain obscure and complex. Even with the introduction of tocolytic therapy, improvements in diagnosis of multiple gestation and placenta previa by ultrasound, as well as government funded nutrition programs, a significant decrease in preterm delivery rates has not occurred for more than two decades. Almost 9% of all neonates in the United States continue to be born before 37 weeks gestation (U.S. Department of Health and Human Services, 1986). Numerous factors associated with the risk of prematurity have been studied, including maternal age and race, education, occupation and working during pregnancy, social class, multiple births, cigarette smoking and substance abuse, and previous preterm birth.

The most disturbing demographics deal with maternal age and race. In the United States, a black woman is at least twice as likely to deliver a live neonate less than 2,500 grams than a white, Cuban, Mexican, Chinese, Japanese, or American Indian woman (U.S. Department of Health and Human Services, 1985). This discrepancy is seen in Figure 9. Reasons for the difference are not understood since the gap remains even after matching for age, parity, education, marital status, perinatal care, nutrition, smoking, and substance abuse (Shiono and Klebanoff, 1986; Kleinman and Kessel, 1987). The racial factor is so extreme that white mothers at the highest risk (less than 15 years of age) have a percentage of low birthweight deliveries similar to black women of ideal child-bearing age (Figure 10). Maternal age carries the highest risks during the teens and at 35 years and above. In the past few decades the delaying of childbearing has increased the population of women having their first child at the higher age of 35+ years (Manitoba Centre for Health Policy and Evaluation Report, 1991).

Here in Manitoba with a substantial Native Indian population, there exists a surprising difference in the incidence of low birthweight deliveries between native and non-native women (Manitoba Centre for Health

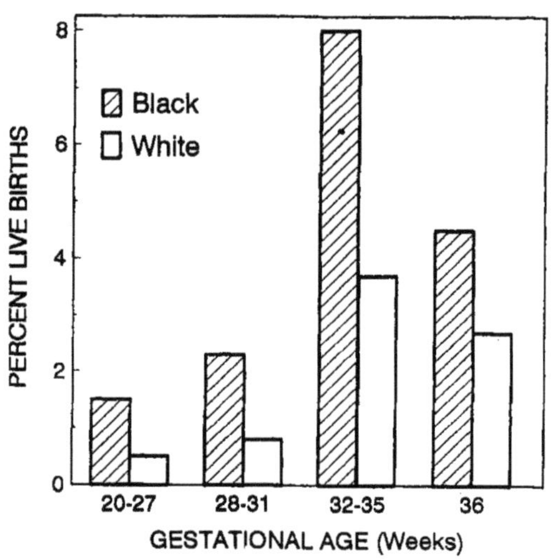

Figure 9. Percentage of U.S. preterm live births within gestational age categories and segregated by race. (Modified from U.S. Department of Health and Human Services, 1985.) See text and Figure 10 for details.

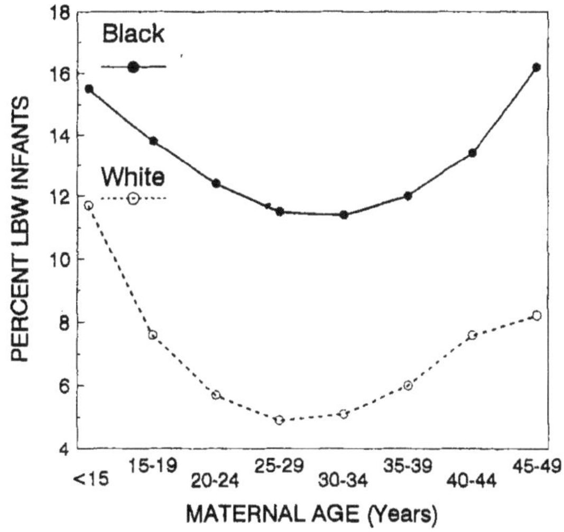

Figure 10. Percentage of U.S. low birthweight (LBW) infants by maternal age categories and segregated by race. (Modified from U.S. Department of Health and Human Services, 1985.)

Policy and Evaluation Report, 1991). The incidence of delivery of low birthweight infants in the native population of first birth mothers less than 20 years of age is 43.3/1,000 live single births as compared to 63.5/1,000 for non-natives (Manitoba Centre for Health Policy and Evaluation Report, 1991). The difference between 20–34 years of age is less but still favors the native women (51.1 vs. 53.7, respectively). For second to fourth births among women aged less than 20 years, the incidences are 52.1 versus 66.3 in favor of native women, but for ages 20–34 the trend reverses to 53.1 versus 45.3, respectively. That finding contradicts the general belief that in all races teenage mothers are more likely to deliver low birthweight neonates than older mothers before the age of 35 years.

Other factors include maternal smoking that decreases birth weight according to the number of cigarettes smoked per day (Wen et al., 1990). The contributions of alcohol consumption and drug abuse to prematurity are as yet unresolved but do relate to lower socioeconomic status that is a factor in premature birth (Lobel et al., 1992). Attempts to determine the effects of maintaining employment during pregnancy have produced contradictory findings. A review of data from 1959–1966 in the United States was unable to show that gestation was shortened in women who worked in the third trimester. However, newborn infants of mothers who

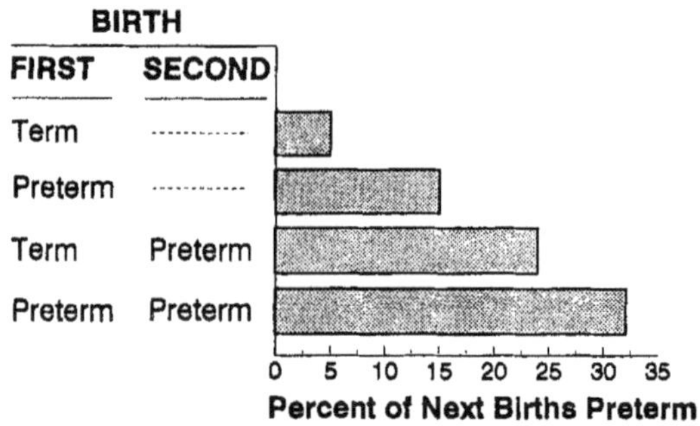

Figure 11. Percentage of births that are preterm following the first or second birth according to whether it was term or preterm. See text for details. (Data modified from Carr-Hill and Hall, 1985.)

did work during pregnancy were found to weigh 150–400 grams less than newborns of mothers who remained at home (Nayey and Peters, 1982). The issue of stress on pregnancy progression and outcome has raised interest because it has a direct relation to many of the other social factors such as smoking, education, social class, marital status, and maternal age that relate to prematurity.

Another important aspect of preterm birth is the relationship between the increased occurrence of a second birth being preterm when it succeeds a previous preterm birth (Figure 11). When the first birth is term, the chances of the next birth being preterm is only 5%. However, when an initial birth is preterm there is a 15% chance that the second birth will be preterm. If the first birth is term and the second birth is preterm, the chance that the next birth will be preterm is 25%. If the first and second births are preterm, the risk increases to a 32% chance that a subsequent birth will also be preterm. Inversely, with every birth that is not preterm, the risk of a subsequent preterm birth decreases (Carr-Hill and Hall, 1985). Early spontaneous or therapeutic abortion appears not to influence the gestational period in subsequent pregnancies.

VIII. HOW IS PREMATURE BIRTH PREVENTED?

As with maternal and infant mortality, prematurity is a function of both environmental conditions and maternal intrinsic factors. Environmental conditions relate mostly to standard of living, education, personal mental and physical hygiene, and availability of health care services. Improvement in those environmental conditions has a positive impact on pregnancy outcome, including prematurity. On the other hand, maternal intrinsic factors relate to underlying pathologies that are manifest even under the best of environmental conditions. However, there is evidence that past practices of therapeutic abortion by rapid and forceful dilatation and curettage increased the risk of prematurity by affecting cervical integrity (Mocsary and Csapo, 1978).

Creasy (1994) suggests that preterm prevention programs have a positive effect on reduction of prematurity in many, but not all, centers in the United States. However, Mustard and Roos (1954), reporting on 12,646 Canadian women, found that while more lower birthweights occurred among those women with the lowest income and prenatal care utilization, the average weight difference as compared to the highest income women was only 58 grams. Moreover, maternal risk factors, including smoking and marital status, were higher among the lowest income group. However, they could find no substantial impact of an already high standard of prenatal care on birthweight in complicated pregnancies. Thus, the impact of universally available prenatal care services on identification and treatment of high risk factors benefit the health of mother and fetus. However, mothers who are noncompliant to prenatal advice jeopardize their pregnancy outcome.

Here at the Women's Hospital in Manitoba, Canada, efforts to even further enhance prenatal care services for pregnant adolescents failed to significantly improve outcome over the already intensive care program currently in place (Drs. Margaret Morris and C. Burckhardt, unpublished observations). This does not argue for reduction in prenatal care services, but rather, it demonstrates the effectiveness of intensive prenatal care leaving maternal noncompliance to prenatal health advice as a risk factor requiring serious attention. Thus, in a society with one of the highest living standards and a universally accessible health care system, we suggest that the manifestation of prematurity is principally due to intrinsic biological reproductive failure. In Canada as of 1989, the incidence of low birthweight was approximately 55/1,000 live births. The theoretical threshold is believed to be 30–35/1,000 (Manitoba Centre for Health Policy and Evaluation Report, 1991). Therefore, to further reduce prematurity below threshold, it is essential to address the underlying nature of the physical causes of labor initiation and progression toward retaining the fetus at risk in its mother's womb until it is able to survive without mechanical and pharmacologic support. Currently, tocolytic therapy to prevent labor and prolong gestation is the strategy of choice for prematurity. Unfortunately, the efficacy of tocolysis in significantly reducing preterm birth is less than expected.

IX. WHAT INITIATES AND SUSTAINS LABOR?

The current state of knowledge of the mechanism of labor is presented in a number of scholarly works on parturition in other chapters in this volume. However, it is becoming clear that the manifestation of parturition commences well in advance of the uterine contractions and cervical dilatation associated with definitive labor. Indeed, there is no clear definition of when parturition actually begins. While we think of labor as an active process, some suggest that it may be the suspension of phase 0 uterine quiescence that initiates labor (Cunningham et al., 1993c). That is, until near the end of pregnancy the state of nonlabor may be an active paralysis of uterine activity as opposed to a passive quiescence awaiting stimulation. Thus, labor onset would not require *de novo* synthesis of biochemical initiator(s) or uterotonins but rather, it may require a suspension of quiescence. Alternatively, the quiescent uterus may require stimulation for the onset and maintenance of contractility. However, outside of pregnancy and even in early pregnancy, the uterus is far from being a quiescent organ. Rhythmic retrograde myometrial contractions have been identified in nonpregnancy, pregnancy up to 10 weeks, and in post-menopausal women suggesting that variable and directional uterine contractions are characteristic of different physiologic states (de Vries et al., 1990).

Studies of the endocrine and chemical interactivity between mother, fetus, and placenta in pregnancy, labor onset, and progression remain incomplete. Good animal evidence exists to suggest a dominant fetal role in initiating the process that originates in the fetal brain, specifically the hypothalamus (McDonald and Nathanielsz, 1991). The fetus as master of its own destiny is readily seen in egg-laying animals, including the few oviparous mammals, and to a great extent in marsupial, bovine, and ovine species. Thus, the concept of fetal control over the timing of the birthing process is not unprecedented in nature. However, while certain fetal abnormalities including anencephaly, fetal adrenal hypoplasia, and placental sulfatase deficiency are associated with the prolongation of gestation or dysfunctional labor, delay of labor onset is not always observed. Moreover, labor onset is known to occur even after fetal death (Cunningham et al., 1993c). Thus, there may be a number of mechanisms from which mother or fetus can draw upon to initiate labor.

Parturition requires uterine responsiveness at the level of both the cervix and the myometrium (Figure 12). Throughout most of gestation the cervix is rigid and the cervical canal is closed with a thick plug of mucus. The myometrium is relatively quiescent but irregular painless contractions do occur from as early as the first trimester. As term approaches, the contractions increase and are particularly apparent in the last one to two weeks of gestation. These non-laboring contractions are called Braxton Hicks contractions and are often responsible for false labor hospital admissions.

Key to successful labor is progressive cervical softening and dilatation that requires the breakdown of collagen, the major component of the

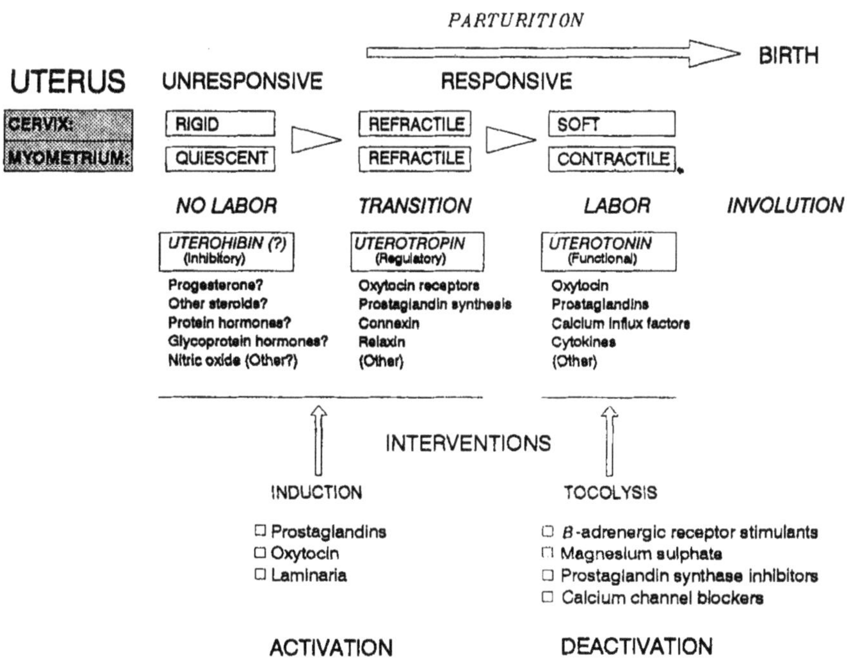

Figure 12. Summary of uterine changes from a rigid cervix and quiescent myometrium to activation of parturition. The model favors an inhibitory influence (Uterohibin?) to maintain uterine unresponsiveness. Characteristic of parturition is when the uterus becomes refractile either by the removal of inhibitory factors or by the appearance of regulatory uterotropin factors. Active labor involves known functional uterotonins. Below are listed interventions that are known to pharmacologically or physically (Laminaria) activate (induction) or pharmacologically deactivate (tocolysis) labor.

cervix. The smooth muscle of the myometrium must also become sensitized from a quiescent to a contractile tissue. The events leading to these changes occur gradually over the last few weeks of pregnancy. Once initiated, a number of compounds and newly generated receptors come into play in the orchestration of labor and delivery. Prostaglandins, oxytocin, relaxin, myometrial gap junction formation, and generation of specific receptors, steroid secretion, cytokine formation, nitric oxide and oxygen radicals, and numerous other factors have been found associated with the process of labor. However, there is as yet no identifiable acute endocrine or biochemical event that precedes labor onset in the human. Thus, that parturition has begun continues to be a call made by the parturient. The major clinical problem with labor onset is when it occurs prematurely at a time when the fetus is developmentally unprepared for birth. The intervention then is to attempt to arrest labor by provoking the contractile myometrium to the quiescent state through tocolytic therapy. In many women, labor, once initiated, appears to be irreversible due to tocolytic resistance. In other women where tocolytic therapy is contraindicated, the birth necessarily proceeds.

X. TOCOLYTIC AGENTS

Tocolytic agents are used to delay delivery in women experiencing premature labor via mechanisms that control or inhibit uterine smooth muscle contractions (Alexander et al., 1993; Creasy, 1994). Tocolysis literally means "to destroy birth." These agents are efficacious in delaying delivery for 1–3 days and may thereby improve the neonate's probability of survival (Graber, 1992; Alexander et al., 1993; Creasy, 1994). This becomes especially important at the crucial stage of 24–27 weeks gestation where delaying delivery by even a few days may benefit the neonate (Wilkins and Creasy, 1990). Nevertheless, controversy remains as to whether tocolytic use has resulted in a consistent decrease in premature births or a decrease in perinatal mortality and morbidity. The most widely used tocolytic agents are the beta adrenergic agonists (beta mimetics) and magnesium sulfate, followed by prostaglandin synthase inhibitors and calcium antagonists or calcium channel blockers, all of which have about equivalent effectiveness (Graber, 1992; Creasy, 1994).

Only 20–30% of women in premature labor are eligible to receive tocolytic therapy. The efficacies of the various tocolytic agents are comparable to each other and depend on factors such as gestational age at the time of tocolytic therapy, amount of cervical dilatation, and degree of effacement, intact or ruptured membranes, and infection (Wilkins and Creasy, 1990; Alexander et al., 1993).

A. Beta Mimetics

Beta-adrenergic agonists operate on myometrial cell membrane receptors to activate adenylate cyclase in the conversion of ATP to cAMP. In turn, the increase in intracellular cAMP activates cAMP-dependent protein kinase that decreases intracellular calcium, leading to reduced myometrial contractility. The beta-adrenergics include isosuprine, hexoprenaline, fenoterol, orciprenaline, ritodrine, salbutamol, and terbutaline (Creasy, 1994). Side effects of the beta mimetics include maternal hypotension, tachycardia, cardiac arrhythmias, myocardial ischemia, chest pain, and pulmonary edema (Wilkins and Creasy, 1990; Beckman et al., 1992; Graber, 1992; Creasy, 1994). Since most of the beta adrenergics cross the placenta, adverse effects on the fetus include mild tachycardia, focal myocardial necrosis, abnormal fetal rhythms, abnormal electrocardiograms, congestive failure, and hydrops (Creasy, 1994). Prolonged exposure to beta-adrenergic agonists leads to downregulation of the receptors and desensitization to treatment (Caritis et al., 1988).

B. Magnesium Sulfate

Magnesium sulfate is proposed to decrease uterine smooth muscle contractility by antagonistically competing with intracellular calcium. Maternal side effects of this tocolytic include flushing, nausea, muscle weakness, chest pain, pulmonary edema, headache, nystagmus, dizziness, lethargy, dryness of the mouth, blurred vision, transient ischemia, urticarial eruption, maternal hypothermia, and neuromuscular blockade. The effects on the fetus include respiratory and motor depression, decreased muscle tone, and drowsiness in the neonate. Prolonged therapy can result in demineralization of long bones and congenital rickets (Wilkins and Creasy, 1990; Beckman et al., 1992; Creasy, 1994).

C. Indomethacin

Evidence suggests that prostaglandins mediate uterine contractions by increasing myometrial gap junctions and intracellular calcium. The prostaglandin synthase inhibitors such as indomethacin, aspirin, naproxen, and fenoprofen decrease uterine contractions by inhibiting activity of the cyclo-oxygenase enzyme that converts arachidonic acid to the various prostaglandins. Maternal side effects with these tocolytics include postpartum hemorrhage and gastrointestinal effects (Creasy, 1994). Prolonged use is associated with headaches, dizziness, depression, psychosis, as well as oligohydramnios which is thought to be a result of decreased urine output by the fetus. Indomethacin readily crosses the placenta and exposure *in utero* may cause pulmonary hypertension in the fetus due to closure of the ductus arteriosus, abnormal cardiopulmonary adaptation in the neonate (Graber, 1992; Creasy, 1994) and growth retardation (Beckman et al., 1992).

D. Calcium Channel Blockers

Calcium channel blockers such as nifedipine are proposed to inhibit intracellular calcium influx through either cell membrane depolarization or receptor interference, and thereby inhibit uterine smooth muscle contractility. Side effects with this tocolytic include vasodilation, flushing, transient headache or nausea, hepatotoxicity (Creasy, 1994), a possible decrease in uteroplacental blood flow, fetal hypoxia, and hypercapnia (Beckman et al., 1992).

E. Antibiotics

There is increasing evidence that ascending infection by pathogenic organisms may involve the cervix and fetal membranes resulting in the release of prostaglandins, uterine contractility, and in some cases, spontaneous premature rupture of membranes. Broad spectrum antibiotic treatment may therefore reduce the risk of labor progressing to delivery. Prophylactic antibiotic administration before labor has yet to be proven effective. Some recent reports suggest treatment efficacy in prolonging gestation in women between 26 and 34 weeks gestation with singleton pregnancies and intact membranes (Norman et al., 1994). Broad spectrum antibiotic treatment following membrane rupture in the same gestational age range is also reported to significantly prolong the latency period (Christmas et al., 1992; Owen et al., 1993). However, Romero et al. (1993) were unable to detect a significant beneficial effect of antibiotics in preterm labor with intact membranes.

F. Nitric Oxide

A new tocolytic agent that is being investigated in the management of premature labor is nitric oxide. Nitric oxide is a potent vasodilator produced by endothelial cells and causes smooth muscle relaxation in not only the vasculature but also in the intestine and the uterus. The onset of labor in animals is associated with decreased nitric oxide synthesis in the uterus. In a recent study, the use of glyceryl trinitrate (a nitric oxide donor) patches applied to the abdomen of 13 women in premature labor at 23–33 weeks gestation, inhibited uterine contractions and prolonged pregnancy by an average of 34 days. There were no cardiorespiratory effects on the fetus or neonate, and maternal side effects associated with this therapy were also minimal, with one-third of patients reporting a headache when more than one patch was applied (Lees et al., 1994). While this tocolytic therapy appears promising, further clinical trials using larger populations are needed to establish its safety and efficacy.

G. Contraindications for Tocolytic Therapy

All the tocolytic agents are associated with serious side effects to both the mother and fetus. Tocolytic therapy is contraindicated for prolonging pregnancy in cases of chorioamnionitis, fetal death, or severe pregnancy induced hypertension (Wilkins and Creasy, 1990). Other contraindications include advanced labor, a mature or anomalous fetus, significant vaginal bleeding (Beckman et al., 1992), premature rupture of membranes, and intrauterine growth retardation (Graber, 1992). Beta-adrenergic therapy is especially contraindicated in cases of known cardiac disease (Creasy, 1994), thyrotoxicosis, and hypertrophic subaortic stenosis (Wilkins and Creasy, 1990), whereas indomethacin therapy is contraindicated in patients with drug-induced asthma, coagulation disorders, hepatic or renal insufficiency, and peptic ulcer disease (Creasy, 1994).

Because tocolytics have a minimum effect in arresting labor, time gained is often used to pharmacologically promote fetal lung maturation. The administration of corticosteroids in patients with intact membranes promotes fetal lung maturation and decreases the incidence of respiratory distress syndrome in neonates less than 34 weeks gestation if a minimum of 24 hours lapses before delivery. Corticosteroid treatment in patients with ruptured membranes has also been shown to decrease the incidence of neonatal respiratory distress syndrome without increasing neonatal infectious morbidity (Wilkins and Creasy, 1990). Moreover, corticosteroid therapy appears to decrease the incidence of intracranial hemorrhage, necrotizing enterocolitis, and neonatal death (Graber, 1992). A common strategy for prematurity now is the combined use of indomethacin for tocolysis and administration of betamethazone to induce fetal lung maturation (Cunningham et al., 1993d).

XI. CONCLUSIONS

Prevention of prematurity requires strategies that address extrinsic and intrinsic deficiencies that provoke the onset of preterm labor. Extrinsic causes relate to living conditions while intrinsic factors relate to pathologies of pregnancy, genetics, and other as yet unknown biological deficiencies. Solutions to the environmental equation are well known and similar to the evolution of the reductions in maternal and infant mortality that accompanied the dramatic upward swing in the socioeconomic development and standard of living in North America and other advanced regions of the world. However, discrepancies between living standards within nations continue as indirectly causative of higher than normal prematurity, and infant and maternal mortality rates.

At the biological level, fundamental research into the mechanism of labor onset and progression is required if the rate of premature childbirth is to be reduced to threshold and below. Current research focuses on the endocrinology and biochemistry of labor, studies on the molecular biology of labor initiation, and the role of maternal infection and disease in preterm labor onset. Because of the numerous etiologies of preterm labor onset, it can be expected that research will identify a variety of new solutions and therapeutic regimens to fit the specific causes. As solutions are defined, interventions will evolve to prolong labor in appropriate cases toward shifting the time of delivery until the fetus reaches a higher weight class compatible with both survival and long-term health.

DEDICATION

Dr. James G. Allardice died suddenly in early 1996. This chapter, which he coauthored, is dedicated to his memory as an exceptional obstetrician and colleague.

ACKNOWLEDGMENTS

We thank Dr. F.A. Manning for availing us of data from the Fetal Assessment Unit that was used in this review. We also appreciate Drs. Cam Mustard and Noralou Roos of the Manitoba Centre for Health Policy and Evaluation for kindly providing information and discussion on pertinent epidemiologic data.

REFERENCES

Alexander, G.R., Weiss, J., Hulsey, T.C., & Papiernik, E. (1993). Preterm birth prevention: An evaluation of programs in the United States. Birth 18, 160–169.

Arias, F., & Tomich, P. (1982). Etiology and outcome of low birthweight and preterm infants. Obstet. Gynecol. 60, 277.

Beckman, C., Ling, F., Barzansky, B., Bates, G.W., Herbert ,W., Laube, D., & Smith, R. (1992). Preterm labor and preterm birth. In: *Obstetrics and Gynecology for Medical Students* (Beckman, C., Ling, F., Barzansky, B., Bates, G.W., Herbert, W., Laube, D., & Smith, R., Eds.) pp. 113–118. Williams and Wilkins, Baltimore, MD.

Brockkov, H., & Kostrava, A. (1973). Sex ratio among human embryos and newborns in a Russian population. Hum. Genet. 17, 91.

Caritis, S.N., Darby, M.J., & Chan, L. (1986). Pharmacologic treatment of preterm labor. Clin. Obstet. Gynecol. 31, 635.

Carr-Hill, R.A., & Hall, M.H. (1985). The repetition of spontaneous preterm labor. Br. J. Obstet. Gynecol. 92, 921.

Christmas, J.T., Cox, S.M., Andrews, W., Dax, J., Leuino, K.J., & Gilstrap, L.C. (1992). Expectant management of preterm ruptured membranes: Effects of antimicrobial therapy. Obstet. Gynecol. 80, 759.

Copper, R.L., Goldenberg, R.L., Creasy, R.K., DeBard, M.B., Davids, R.O., Entman, S.S., Iams, Y.D., & Cliver, S.P. (1993). A multicenter study of preterm birth weight and gestational age specific mortality. Am. J. Obstet. Gynecol. 168, 78.

Creasy, R.K. (1994). Preterm labor and delivery. In: *Maternal Fetal Medicine: Principles and Practice. Part III. Maternal and Fetal Pathophysiology. Disorders of Parturition* (Creasy, R.K., & Resnik. R., Eds.) pp. 477–504. W.B. Saunders Company, Philadelphia, PA.

Cunningham, F.G., MacDonald, P.C., Gant, N.F., Leveno, K.J., & Gilstrap, L.C. (1993a). Obstetrics in broad perspective. In: *Williams Obstetrics*, 19th ed. (Cunningham, F.G., MacDonald, P.C., Gant, N.F., Leveno, K.J., & Gilstrap, L.C., Eds.) pp. 1–9. Appleton and Lange, Norwalk, CT.

Cunningham, F.G., MacDonald, P.C., Gant, N.F., Leveno, K.J., & Gilstrap, L.C. (1993b). Abortion. In: *Williams Obstetrics*, 19th ed. (Cunningham, F.G., MacDonald, P.C., Gant, N.F., Leveno, K.J., & Gilstrap, L.C., Eds.) pp. 661–690. Appleton and Lange, Norwalk, CT.

Cunningham, F.G., MacDonald, P.C., Gant, N.F., Leveno, K.J., & Gilstrap, L.C. (1993c). Parturition: Biomolecular and physiologic processes. In: *Williams Obstetrics*, 19th ed. (Cunningham, F.G., MacDonald, P.C., Gant, N.F., Leveno, K.J., & Gilstrap, L.C., Eds.) pp. 297–361. Appleton and Lange, Norwalk, CT.

Cunningham, F.G., MacDonald, P.C., Gant, N.F., Leveno, K.J., & Gilstrap, L.C. (1993d). Preterm and postterm pregnancy and fetal growth retardation. In: *Williams Obstetrics*, 19th ed. (Cunningham, F.G., MacDonald, P.C., Gant, N.F., Leveno, K.J., & Gilstrap, L.C., Eds.) pp. 853–889. Appleton and Lange, Norwalk, CT.

de Vries, K., Lyons, E.A., Ballard, G.,Levi, C.S., & Lindsay, D.Y. (1990). Contractions of the inner third of the myometrium. Am. J. Obstet. Gynecol. 162, 679.

Demographic Yearbook (U.N.). (1992). In: *The Canadian Global Almanac, 1993* (Colombo, J.R., Ed.). Macmillan Canada, Toronto.

Gardosi, J., Chang, A., Kalyan, B., Sahota, D., & Symonds, E.M. (1992). Customized antenatal growth charts. Lancet 339, 283.

Graber, E.A. (1992). Prematurity 1992. Obstet. Gynecol. Survey 47, 521–524.

Graves, E.J. (1993). *1991 Summary: National Hospital Discharge Survey*. U.S. Department of Health and Human Services, Hyattsville, MD. CDC Advance data no. 227.

Gruenwald, P. (1963). Chronic fetal distress and placental insufficiency. Biol. Neonate 5, 215.

Harlap, S., Shiono, P.H., & Ramcharan, S. (1980). A life table of spontaneous abortions and the effect of age, parity and other variables. In: *Human Embryonic and Fetal Death* (Porter, I.H., & Hook, E.B., Eds.) pp. 145–149. Academic Press, New York.

Hytten, F.E. (1982). Boys and girls. Br. J. Obstet. Gynecol. 89, 97.

Institute of Medicine. (1985). *Preventing Low Birthweight*. National Academy Press, Washington, DC.

Kellakumpu-Lehtinen, P., & Pelliniemi, L. (1984). Sex ratio of human conceptuses. Obstet. Gynecol. 64, 220.

Kleinman, J.C., & Kessel S.S. (1987). Racial differences in low birth weight: Trends and risk factors. N. Engl. J. Med. 317, 749.

Lees, C., Campbell, S., Jauniaux, E., Brown, R., Ramsay, B., Gibb, D., Moncada, S., & Martin, J.F. (1994). Arrest of preterm labor and prolongation of gestation with glyceryl trinitrate, a nitric oxide donor. Lancet 343, 1325–1326.

Lobel, M., Dunkel-Schetter, C., & Scrimshaw, P.C. (1992). Prenatal maternal stress and prematurity. A prospective study of socioeconomically disadvantaged women. Health Psychol. 11, 32.

Manitoba Centre for Health Policy and Evaluation Report #91-II-04. (1991). Maternal Demographic Risk Factors and the Incidence of Low Birthweight, Manitoba 1979–1989.

Manitoba Centre for Health Policy and Evaluation Report #93-01. (1993). The Utilization of Prenatal Care and Relationship to Birthweight Outcome in Winnipeg, 1987–1988

McDonald, T.J., & Nathanielsz, P.W. (1991). Bilateral destruction of the fetal paraventricular nuclei prolongs gestation in sheep. Am. J. Obstet. Gynecol. 165, 764.

Mocsary, P., & Csapo, Al. (1978). Effect of menstrual induction on prematurity rate. Lancet 2, 1159.

Morales, W.J. (1987). The effect of chorioamnionitis on the developmental outcome of preterm infants at one year. Obstet. Gynecol. 70, 183.

Mustard, C.A., & Roos, N.P. (1954). The relationship of prenatal care and pregnancy complications to birthweight in Winnipeg, Canada. Am. J. Public Health 84, 1450.

Nayey, R.L., & Peters, E.C. (1982). Working during pregnancy: Effects on the fetus. Pediatrics 69, 724.

Norman, K., Pattinson, R.C., de Souza, J., DeJong, P., Moller, G., & Kirsten, G. (1994). Ampicillin and metronidazole treatment in preterm labor: A multicentre, randomized controlled trial. Br. J. Obstet. Gynecol. 101, 404.

Owen, J., Groome, L.J., & Hauth, J.C. (1993). Randomized trial of prophylactic antibiotic therapy after preterm amnion rupture. Am. J. Obstet. Gynecol. 169, 976.

Romero, R., & Major, M. (1988). Infection and preterm labor. Clin. Obstet. Gynecol. 32, 553.

Romero, R., Sibai, B., Caritis, S., Paul, R., Depp, R., Rosen, M., Klebanoff, M., Sabo, V., Evans, J., Thom, E., Cefalo, R., & McNellis, D. (1993). Antibiotic treatment of preterm labor with intact membranes: A multicenter, randomized, double-blinded, placebo- controlled trial. Am. J. Obstet. Gynecol. 169, 764.

Savity, D.A., Blackmore, C.H., & Thorp, J.M. (1991). Epidemiologic characteristics of preterm delivery: Etiologic heterogeneity. Am. J. Obstet. Gynecol. 164, 467.

Schwartz, R.M. (1989). What price prematurity? Family Planning Prospect 21, 170.

Shiono, P.H., & Klebanoff, M.A. (1986). Ethnic differences in very preterm delivery. Am. J. Public Health 76, 1317.

Spreet, H. (1980). *Obstetrics and Gynecology in America: A History.* Waverly Press, Baltimore, MD.

Steinberg, W.M. (1989). Maternal mortality. Curr. Opin. Obstet. Gynecol. 1, 137.

United States Bureau of the Census, World Population Profile. (1994). U.S. Public Health Service, Washington, DC.

U.S. Department of Health and Human Services. (1985). Report of the Secretary's Task Force on Black and Minority Health, Publication 0–487–637 (QL3). Vol. 6, Infant mortality and low birthweight. National Center for Health Statistics, Hyattsville, MD.

U.S. Department of Health and Human Services. (1986). Public Health Service, Vital Statistics of the United States, 1982. Vol. I, Natality. National Center for Health Statistics, Hyattsville, MD.

Walker D.Y.B., Feldman, A., Vohr, B.R., & Oh,W. (1984). Cost-benefit analysis of neonatal intensive care for infants weighing less than 1,000 grams at birth. Pediatrics 74, 20.

Wen, S.W., Goldenberg, R.L., Cutter, G.R., Hoffman, H.Y., Cliver, S.P., Davis, R.O., & DuBard, M.B. (1990). Smoking, maternal age, fetal growth and gestational age at delivery. Am. J. Obstet. Gynecol. 162, 53

Wilkins, I., & Creasy, R.K. (1990). Preterm Labor. Clin. Obstet. Gynecol. 33, 502.

World Health Organization. (1992). Reproductive Health, A Key to a Brighter Future, Geneva.

World Health Statistics Annual, 1993. (1994). World Health Organization. Geneva.

Wright, J.W. (1994). *The Universal Almanac: 1995* (Wright, J.W., Ed.). Andrews and McMeel, Kansis City, MI.

INDEX

www.ingramcontent.com/pod-product-compliance
Ingram Content Group UK Ltd.
Pitfield, Milton Keynes, MK11 3LW, UK
UKHW020808290926
15470UKWH00003B/1036

9 781559 386395